recent advances in phytochemistry

volume 16

Cellular and Subcellular Localization in Plant Metabolism

RECENT ADVANCES IN PHYTOCHEMISTRY

Proceedings of the Phytochemical Society of North America
General Editor: **Frank A. Loewus** *Washington State University, Pullman, Washington*

Recent Volumes in the Series

recent advances in phytochemistry

volume 16

Cellular and Subcellular Localization in Plant Metabolism

Edited by

Leroy L. Creasy
Cornell University
Ithaca, New York

and

Geza Hrazdina
New York State Agricultural Experiment Station
Geneva, New York

PLENUM PRESS • NEW YORK AND LONDON

Library of Congress Cataloging in Publication Data

Main entry under title:

Cellular and subcellular localization in plant metabolism.

(Recent advances in phytochemistry; v. 16)
Papers presented at a Symposium on Cellular and Subcellular Localization in
Plant Metabolism during the annual meeting of the Phytochemical Society of North
America, at Cornell University, Ithaca, N.Y., on Aug. 10 – 14, 1981.
Bibliography: p.
Includes index.
1. Plants—Metabolism—Congresses. 2. Plant cells and tissues—Congresses. I.
Creasy, Leroy L. II. Hrazdina. G. III. Symposium on Cellular and Subcellular Lo-
calization in Plant Metabolism (1981: Cornell University). IV.   Phytochemical
Society of North America. Meeting (21st: 1981: Cornell University) V. Series.
QK861.R38   vol. 16 581.19'2s [581.1'33]   82-7560
[QK881]   AACR2
ISBN 0-306-41023-0

Proceedings of the Twenty-first Annual Meeting of the Phytochemical
Society of North America, held August 10 – 14, 1981, at Cornell
University, Ithaca, New York

©1982 Plenum Press, New York
A Division of Plenum Publishing Corporation
233 Spring Street, New York, N.Y. 10013

Printed in the United States of America

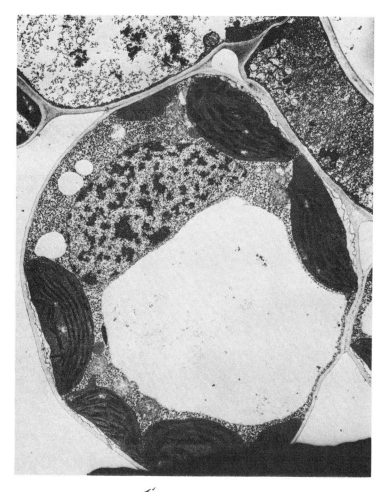

A mesophyll cell from a leaf of _Phleum pratense_ (5,300 x).
A prominent vacuole, limited by its tonoplast, occupies a
large portion of the cell area in this section. Within the
vacuole are traces of flocculent material of unknown composition. Other recognizable organelles include the nucleus,
flattened where it is in proximity to the central vacuole,
plastids, small vacuoles, mitochondria, microbodies, ribo-
somes and a scant amount of endoplasmic reticulum. The
plasma membrane limits the protoplast and to its exterior
is a thin cell wall. The wall is in contact with other
cells in some places and with intercellular gas spaces at
others. The dark irregular boundary in this micrograph is
a portion of a grid bar. Myron C. Ledbetter, Brookhaven
National Laboratory.

PREFACE

Morphological differences between cells and the existence of morphologically distinct particles have been examined since cells were first recognized. Each technological advance in detection and visualization has led to the description of different organelles and cell types. Basic biochemical processes in cells were recognized and are now well understood. It is only recently however, that research has expanded to include the specific metabolic function of the specialized cell types and organelles. In some cases metabolic roles were recognized when the organelles were first described, e.g., chloroplasts, mitochondria, etc., in others the metabolic role remains unknown. Chemical and biochemical specialization in plants or their organelles is equally challenging. Although biochemists have laboured intensively on many isolated plant organelles, it is only recently that technical advances have permitted the examination of specialization in the metabolism of cell types. This area of research, although under intensive investigation in some areas of plant metabolism, is still in its infancy. Further developments in methodology or in production of specific genetic lines of plants will greatly improve our understanding of the specialization of different tissues and cell types.

This volume describes the current status in the discipline as presented in a Symposium on the Cellular and Subcellular Specialization in Plant Metabolism during the Annual Meeting of the Phytochemical Society of North America, at Cornell University, Ithaca, N.Y., on August 10-14, 1981. The authors of the individual chapters were specifically selected by the Society for their potential in the new area of cellular and subcellular metabolic specialization. They represent the new wave of recognized experts in this field and are those who will continue to lead. We have enjoyed working with them and acknowledge the efforts expanded both in the oral presentations at the Annual Meeting and in the preparation of the exceptional chapters which follow.

The Symposium could not have been realized without the financial contribution of the New York State College of Agriculture and Life Sciences, Cornell University, and the Conversations in the Disciplines Program of the State University of New York.

January, 1982

L. L. Creasy
G. Hrazdina

CONTENTS

Chapter One

COMPARTMENTATION IN PLANT CELLS: THE ROLE OF THE VACUOLE

GEORGE J. WAGNER

Biology Department
Brookhaven National Laboratory
Upton, New York   11973

INTRODUCTION

Roles of the Vacuole

The mature plant cell vacuole is a multifunctional
organelle which is unique to higher plants. It constitutes
a compartment which is thought to be anabolically inactive
and a primary site for metabolite storage and sequestration.[1,2] Cytologists carried out extensive studies of
plant vacuoles during the late 1800's and early 1900's
which revealed many of the functions of this organelle.
This chapter will consider the roles of the vacuole in
solute storage and sequestration, and discuss methods now
being used to isolate and investigate plant vacuoles.
Recent studies which have utilized vacuoles to estimate the
compartmentation of solutes and enzymes will be discussed.
Possible mechanisms of tonoplast transport will be
considered.

The primary functions of the mature plant cell vacuole
which are recognized at present are listed in Table 1. The
first four functions listed have been recognized for some
time. The reader is referred to reviews by Guilliermond,[2]

1

Table 1. Principal roles of the mature-plant-cell vacuole.

---

1) Osmotic - In concert with the cytoplasm and cell wall

        a) Mechanical support
        b) Tissue movement
        c) Motive force for cell expansion
        d) Stomatal function

2) Ion balance and storage

3) Metabolite storage

4) Metabolite sequestration

5) Lytic functions, senescence

6) Intracellular and intercellular mixing via transvacuolar strands and plasmodesmata

7) Minimalization of the volume of the cytoplasm,
maximalization of the cytosol-tonoplast interface and
provision for efficient distribution of the photosynthetic apparatus.

Zirkel,[1] Kramer,[3] Voller,[4] DeRobertis et al.,[5] Pisek,[6]
Matile,[7] and Marty et al.[8] for discussion of these
functions as well as for history and general discussion.

    The role of the vacuole in lytic processes has received
considerable attention since methods for efficient isolation
of intact higher plant vacuoles were developed in 1975-76.
Methods for preparing vacuoles from protoplasts have been
particularly useful for these studies. Much of the recent
interest in the lytic function of the vacuole has centered
on its role in proteolysis during chloroplast senescence.
New information on this subject has been discussed in
several recent reports.[8,15] Whether degradation of chloroplast proteins is catalyzed by chloroplast, vacuolar, or
cytoplasmic proteases, or by cooperative action of some or
all of these is a matter of some controversy. Many investigators have shown that the bulk of acid protease and

other acid hydrolases occur within the vacuole of protoplasts prepared from various tissues and plants. However,
quantitative methods used in vacuole/extravacuole localization studies are not sufficiently accurate to rule out the
existence of perhaps as much as 15 percent of the cell acid
hydrolase outside the vacuole (see later discussion and ref.
16). A detailed understanding of the part played by the
mature plant cell vacuole in lytic functions during and
prior to senescence must await further study.

The role of the vacuole in facilitating intracellular
and intercellular mixing via transvacuolar strands is one
which is not often discussed. It is not known if tonoplast
constituents participate in cyclosis through transvacuolar
strands and in directing strand formation or if they play
a passive role in these events. It is clear, however, that
the strand system is important in intracellular communication and, through plasmodesmata, intercellular communication. These processes are emphasized here because the petal
protoplasts we have used in our studies contain few plastids,
and transvacuolar strands and cyclosis through them are
easily recognized.[17] In contrast, these processes are
difficult to observe in "healthy" leaf mesophyll protoplasts
because chloroplasts are concentrated in the peripheral
cytoplasmic layer of such cells and obscure the view of the
interior of the cell. Protoplasts which are left too long
in the enzymic medium used for isolation or those stored
under conditions eventually leading to cell death are
characterized by an aggregate or cap of internal organelles
and membranes at one end of the cell. When this has
occurred fragile transvacuolar strands are disrupted and
cannot be observed. The nucleus often occurs in the center
of a healthy petal protoplast and a transvacuolar strand
network emanates 360° from the nucleus to the periphery of
the protoplast. Cyclosis is observed to be directional
within the strands and cytosol flow is rapid, particularly
in and around the nucleus. The transvacuolar strand network not only facilitates cytoplasmic mixing but also
provides for a large tonoplast-cytosol interface. The
extensibility of the tonoplast of isolated petal vacuoles
has been briefly discussed.[18]

The last function listed in Table 1 has been discussed
in interesting articles by Wiebe,[19] and Walter and
Stadelman[20] and also by Dainty.[21] The large central

vacuole of the plant cell is not essential for satisfying
the osmotic requirements of terrestrial plants,[21] but it
does satisfy the need for a large cell surface (and therefore plant surface) without requiring the maintenance of a
large volume of cytosol filled with energy-expensive solute.
The peripheral cytosol layer permits the chloroplasts to
be distributed close to the cell surface maximizing penetration of light to the photosynthetic apparatus.

Many of the functions listed in Table 1 involve uni-
or bi-directional movement across the tonoplast of solutes
which originate in the cytosol, in or on cytoplasmic membranes or organelles, or come from outside the cell and
traverse the cytosol enroute to the vacuole. Biosynthesis
in the normally acidic vacuole sap is unlikely and has not
been demonstrated. The association of biosynthetic enzymes
with isolated tonoplast has not been reported, although it
has not been ruled out.[18,22] It is clear that the osmotic,
storage, and sequestrative functions of the vacuole are
interrelated in that all osmotically active solutes in the
vacuole are involved in turgor maintenance and other
osmotic functions.

Accumulation in Plant Cells

A characteristic of higher plants which distinguishes
them from other biological organisms is that they form and
accumulate a host of natural products, most of which are
not primary metabolites. These substances often accumulate
within the cell and are known or thought to be confined to
the vacuole sap. The best known of the accumulated metabolites form the chemically heterogeneous group generally
called the secondary plant products. Included in this
group are the alkaloids, cyanogenic glycosides, flavonoids,
and terpenoids. These compounds generally have limited
distribution in the plant kingdom and have been considered
waste products of metabolism.[23,24] Emerging evidence suggests distinct roles for certain secondary plant products in
plant and animal ecology.[25-28] Among the secondary plant
products are the principles which confer many flavors,
textures, colors and odors to our foods, pharmacologically
active substances used to maintain human and animal health,
and pigments which beautify our natural environment and
serve as attractants for pollinators. These substances
greatly enrich the quality of our lives.

A large number of natural products that are accumulated
in certain plants are not always classified as secondary
plant products, e.g., ions, organic acids, sugars, amino
acids, phytotoxic proteins, and other protein derivatives.
These may be stored or sequestered in the vacuole or may
have other pools within cells. For example, oxalic acid
can occur within and outside the cell[29] and malic acid may
exist in three or more pools within the cell.[30-32]

The major groups of natural products which accumulate
in plants and representatives of each group are shown in
Table 2. Examples selected are for the most part extreme
cases, chosen to point out the accumulation potential of
plants. Higher plants also accumulate substances taken up
from the environment. These include trace elements, inorganic salts, and pesticide residues (Table 3). The subcellular sites of accumulation of exogenously derived substances are not well characterized. Some evidence exists
for the vacuolar accumulation of nitrate[50] and the accumulation of certain heavy metals in the cytosol.[51]

PROBLEMS AND PROMISE OF VACUOLE/EXTRAVACUOLE LOCALIZATION STUDIES

With the recent development of methods for isolating
mature plant cell vacuoles in large numbers has come the
means for directly determining the solute composition of
this organelle. By quantitative comparison of the solute
composition of vacuoles and protoplasts, or vacuoles and
tissues, it is possible to estimate the distribution of
solutes between the vacuolar and extravacuolar compartments.
There are certain difficulties in making such comparisons,
which will be discussed in this section. Despite existing
problems, much important information has already emerged
regarding the subcellular location of various stored and
sequestered metabolities. These results will be reviewed
in a subsequent section.

In 1975-1976 methods were described for the isolation
of intact higher plant vacuoles in large numbers.[16,52-55]
These techniques have been adapted for use in estimating
the vacuolar/extravacuolar distribution of a variety of
solutes in different plant systems. In most studies, true
subcellular localization has not been achieved because
comparative analysis of protoplasts and vacuoles compares

Table 2. Examples of natural products which accumulate in
higher plant tissues.

Class of substance	Example	Genus	Maximum level observed (% dry wt.)		Reference
Alkaloid	Nicotine	Nicotiana	5	leaves	33
Cyanogenic glycoside	Dhurrin	Sorghum	30	shoot	34
Flavonoid	Anthocyanin	many genera	30	petal	35
	Rutin	Fagopyrum	8.5	leaves	36,37
Isoprenoid	Digitoxin	Digitalis	0.2	leaves	38
	Diosgenin	Dioscorea	0.7	leaves	39
Phenolic	Tannin	Quercus	5	leaves	40
	2-Hydroxy-cinnamic acid	Melilotus	6.2	plant	41
Aliphatic acid	Oxalate	Halogeton	30	plant	42
		Spinacia	0.3	plant	42
	Malonic acid	Medicago	2.9	plant	41
Phytotoxin	Soybean agglutinin	Glycine	3% of seed protein		43
Amino acid	Glutamine	Dactylis	1.9	plant	41
	γ-Amino-butyric acid	Medicago	0.34	leaves	41
Saccharide	Sucrose	Dactylis	5.8	plant	41
	Ascorbate	Diospyros	13	leaves	44

vacuolar contents with the contents of all extravacuolar
compartments - unless protoplasts are lysed under conditions

Table 3. Examples of accumulated substances which are
derived from the environment.

Class of substance	Example	Genus	Maximum level observed (% dry wt.)		Reference
Inorganic ion	Nitrate	Avena, Zea, Sorghum	25	plant	45
Elements	Se	Astragalus	0.2	plant	46
	Ni	Siebertia	1.1	leaves	47
	Cd	Lycopersicon	0.1	leaves	48
Pesticide residues	DDT	Nicotiana	0.13	plant	49
	Endosulfan	Nicotiana	0.03	plant	49

which preserve the integrity of cytoplasmic organelles.
In a few cases[13,15,56] vacuole/extravacuole distribution
studies have included direct examination of chloroplasts
and have therefore refined the analysis to include this
organelle.

Two general methods have been used to study vacuole/
extravacuole distribution after isolation of vacuoles. In
one of these, protoplasts and vacuoles are prepared and the
number of structures in aliquots of the preparations are
counted in the light microscope using a volume-calibrated
counting chamber. The structures are then lysed and the
concentrations of the solute of interest in the preparations
determined. The vacuole/extravacuole distribution of the
solute in protoplasts is determined by relating solute per
protoplast with solute per vacuole. The second general
method uses a vacuole marker to compare the solute content
of isolated vacuoles with that in tissue or protoplasts.
Leigh et al.[57] used the endogenous vacuolar pigment of red
beet roots, betanin, to relate the sucrose, acid invertase,
and acid phosphatase content of isolated vacuoles to
vacuoles in tissue. The amount of these constituents per
mg of protein was determined for each fraction recovered
during isolation of vacuoles and compared with the betanin

per mg of protein observed in these fractions. Correlation
coefficients close to unity were obtained in these analyses
indicating that like betanin these constituents were primarily vacuolar in this tissue. Grob and Matile[58] assumed
that the phenolics of horseradish root are entirely confined
to the vacuole and compared the yield of phenolics and
ascorbic acid recovered with vacuoles to assess the vacuole/
extravacuole distribution of the constituent. Buser and
Matile[59] introduced the vital stain neutral red into protoplasts of Bryophyllum prior to isolating vacuoles in order
to facilitate quantitation of structures. The vacuole/
extravacuolar distribution of malic acid was determined by
relating malic acid per unit of neutral red of protoplast
and vacuole preparations. Boller and Kende[60] and Martinoia
et al.[50] used a combination of the two methods described.
The α-mannosidase content of protoplasts and vacuoles were
compared by the method based on counting of structures and
it was concluded that all the α-mannosidase of the tissues
examined was contained in the vacuole. The vacuole/extra-
vacuole distribution of various enzymes and protein[50,60]
and amino acid and nitrate[50] were assessed by quantitatively
relating their occurrence to that of α-mannosidase.

Both basic methods just described are imperfect for
different reasons. Current problems in achieving vacuole/
extravacuole distribution analysis and possible solutions
to these problems listed in Table 4 are discussed below:

Methods which utilize protoplasts suffer from two
major difficulties.

1) Changes in metabolism and solute compartmentation
    may occur during protoplast and vacuole isolation.

2) Procedures generally used to determine protoplast
    and vacuole numbers and volumes are inaccurate
    and can lead to errors in the assessment of
    vacuole/extravacuole distribution.

Taylor and Hall[61] and Hall[62] in studying the effects
of plasmolysis and cell-wall-degrading enzymes on the
permeability of tissues prelabeled with ^{32}P and ^{86}Rb,
observed time-dependent release of ^{86}Rb in plasmolyzed
tissue which was accelerated on addition of wall-degrading
enzyme. Release of ^{32}P only occurred on addition of

Table 4. Current problems in vacuole/extravacuole distribution analysis and possible solutions.

Method	Problems	Possible Solutions
Analysis of protoplasts	1) Changes in metabolism and permeability during protoplast isolation	a) Minimize time of protoplast isolation b) Better isolation-enzyme systems
	2) Quantitation	a) Better methods for determination of structural volume b) Use of appropriate vacuole markers
Analysis of tissues	1) Limited to storage root tissues?	New methods for leaves, etc.
	2) Low yields	Improved methods
	3) Recognition of suitable markers	Use of secondary products

wall-degrading enzyme. It was concluded that wall-degrading
enzyme damaged the plasma membrane of cells and caused
leakage of ions from the cytosol. It is not known if loss
of ions from the cytosol results in leakage from the
vacuole. It is unlikely that the tonoplast is damaged by
wall-degrading enzymes since it is not in contact with the
protoplast isolation medium. A number of investigations
have reported that the concentration of various solutes in
isolated vacuoles is identical to that in the protoplasts
from which they were isolated. This suggests that the permeability of the tonoplast is retained in isolated protoplasts. Improved methods for rapid isolation of protoplasts,
which have resulted from the availability of more efficient
wall-degrading enzymes such as Pectinolyase Y-23[63] and
Pectinase[16] (Sigma Chemical Co.), have reduced the time
required to isolate protoplasts from certain tissues.
Presumably the use of such methods minimizes the effects
of wall-degrading enzymes on protoplast permeability and
metabolism. Galun has recently reviewed available information concerning the physiological state of isolated plant
protoplasts.[69a]

The second major difficulty in assessing vacuole/extra-
vacuole solute distribution in protoplasts, a problem in
quantitation, is due to size reduction of protoplasts and
vacuoles which occurs during isolation and purification.
Vacuoles, and to a lesser extent protoplasts, subdivide and
give rise to populations comprised of structures varying in
size. Light micrographs of protoplasts and vacuoles presented by various authors who have used different methods
of preparation show wide variations in structure and size.
The generally applied method in vacuole/extravacuole distribution analysis based on counting structures depends on
visual averaging of structure size variation and estimating
a mean diameter for structures in a population. This is
difficult to achieve rapidly and accurately, particularly
for vacuoles which are fragile and need to be processed
rapidly. In studies of metabolite partitioning the crucial
parameter is the volume and not the number of structures.
Since the volume of isolated protoplasts and vacuoles is a
function of their radius cubed, a poor estimate of the mean
diameter of structures in a population can lead to major
errors in assessing the vacuole/extravacuole distribution
of solutes. Wagner et al.[64] have discussed this problem
and described an improved method of sap volume

determination based on the direct measurement of the diameters of images of protoplasts and vacuoles on photographs
taken during structure preparation. This method can be applied to protoplasts and vacuoles from all tissues, requires
very little sample for quantitation, is rapid, and is little
more tedious than the commonly used counting procedure.
Distribution profiles which describe the variation in protoplast and vacuole size and sap volume in preparations from
Hippeastrum petals and wheat leaves are shown in Figure 1.
Vacuoles subdivide during manipulation more easily than
protoplasts and it is clear than one cannot equate one
protoplast with one vacuole. The degree of subdivision of
protoplasts and vacuoles may vary according to methods of
preparation, purification and the plant source but
structure size variation undoubtedly occurs in all cases
and is most extensive in vacuole preparations.

The use of vacuole markers for quantitation requires
that it be demonstrated that the marker is exclusively
confined to the vacuole in the system being studied. Where
this analysis has been made[50,59] the vacuole and protoplast
content of the marker was compared after counting structures.
The same problems discussed above apply in this case.

Procedures for estimating vacuole/extravacuole solute
distribution in tissues after isolation of vacuoles directly
from tissue also have shortcomings: 1) Mechanical methods
for vacuole isolation may be limited to storage roots and
2) Mechanical isolation methods give low yields. A major
advantage of these procedures, however, is that changes in
metabolism and cell organelle permeability are minimized or
avoided entirely. Their greatest disadvantage is that current methods have been successfuly applied only to storage
root tissues. It is not known if the spacial arrangement
of cells in these tissues, the nature of their cell walls,
or the composition of storage root tonoplast and vacuoles
make this type of tissues suitable for mechanical vacuole
isolation. Another disadvantage of these methods is that
the recovery of intact vacuoles is relatively low. This
makes these methods less suitable for experiments in which
the subcellular location of radio-labeled metabolites or
products of labeled precursors is being studied. Quantitative comparison of the solute content of mechanically
isolated vacuoles and vacuoles in tissue requires the use of
vacuolar sap markers such as betanin in red beet root[57] and

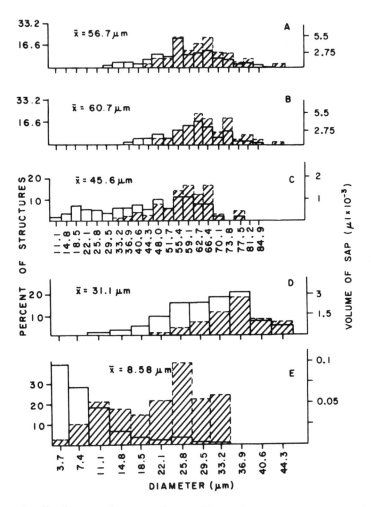

Figure 1. Number and sap volume distributions of protoplasts
and vacuoles in preparations from Hippeastrum and wheat.
Protoplasts and vacuoles were prepared and quantitated as
previously described.[64,67] Open histograms represent the
percent of the total number of structures in a preparation
which was present in each size group (size groups as actual
diameters). Shaded histograms represent the total volume
of sap in structures in each size group calculated from the
number of structures in a group and the mean diameter of
the group. Data are for one experiment for each of (A)

phenolics in horseradish root.[58] Pigments and other secondary plant products are good candidates for vacuole markers.
It is difficult to establish experimentally that markers
are confined to the vacuole in tissue. The method used by
Grob and Matile[58] for this purpose is an interesting one.
In several experiments, they compared the ratios of acid
phosphatase-to-ascorbate and phenolics-to-ascorbate
recovered with mechanically isolated vacuoles of horseradish
root. The ratios were found to be independent of the yield
of vacuoles which suggests that ascorbate like the markers
was primarily confined to the vacuole in this tissue.
Mechanically prepared root vacuoles also vary in size (see
(Figure 7b, reference 53). Therefore, counting vacuoles
for the purpose of quantitating sap volume (or membrane
area) can lead to the same errors discussed above for
applying this method to protoplast-derived vacuoles.

RESULTS OF RECENT STUDIES

     Results of recent efforts to determine the vacuole/
extravacuole distribution of various solutes after isolation
and analysis of vacuoles are listed in Table 5. The amino
acid content of isolated vacuoles has been studied with
three tissues. In Hippeastrum and tulip[65] the percent of
protoplast free amino acid found in vacuoles varied from
leaf to petal (tulip) and from one amino acid to the other.
The non-protein amino acid γ-aminobutyric acid was found to
be primarily extravacuolar in all three tissues. Martinoia
et al.[50] and Heck et al.[13] obtained somewhat different
results for the distribution of total free amino acid in
barley mesophyll protoplasts, but both observed that the
vacuole is a major pool for these substances in the tissue.
The vacuole is thought to be a relatively inactive amino
acid pool in plants[71,72] and references therein. The
vacuole was shown to be a pool for sucrose in four tissues,[57,65,66] and glucose and fructose were observed in
vacuoles of Hippeastrum and tulip tissue.[65] High

Figure 1. (Legend continued) flotation-prepared Hippeastrum
petal protoplasts; (B) washout-prepared Hippeastrum petal
protoplasts; (C) Hippeastrum petal vacuoles; (D) washout-
prepared wheat leaf protoplasts; (E) wheat leaf vacuoles.
Reproduced by permission of the American Society of Plant
Physiologists from reference 64.

Table 5.   Vacuole/extravacuole partitioning of plant cell
metabolites as determined from studies of isolated vacuoles.

Substance	Percent in vacuole	Tissue	Reference
Amino acids	50;52;85	Hippeastrum, tulip; barley; barley	67,50,13
Sucrose	44-100; most; 62	Hippeastrum, tulip; beet root; castor bean	65,57,66
Glucose, fructose	50-100	Hippeastrum, tulip	65
Malic acid	most	Bryophyllum	59
Ascorbate-derived oxalate	70	barley	67
Na^+, Mg^+, Ca^{++}, Cl^+, K^+	most	Hippeastrum, tulip	68
Anthocyanin	100	Hippeastrum, tulip	65
Betanin	all	beet root	57
Ascorbate	100	horseradish root	58
Phenolics	100	horseradish root	
Glucosinolates	100	horseradish root	
Dhurrin	all	Sorghum	69
Nicotine	93	tobacco	33
Proteinase inhibitor	all	tomato	70
Gibberellin	30-100	barley, cowpea	56
Nitrate	99	barley	50

concentrations of the last two sugars have been observed in
vacuole from fruit subepidermal cells of grape berries.[75]
Evidence supporting the accumulation of sugars in a vacuolar
pool has been presented.[30,73,74]

Organic acids are known to accumulate in plant vacuoles
and can often be observed in crystalline form in the vacuole

sap. The bulk of the malic acid in Byrophyllum appears to be confined to the vacuole.[59] Large amounts of organic acid have also been observed in the vacuolar sap of Hippeastrum,[65] grape berries,[75] and Sedum.[76] In Sedum the malic acid content of isolated vacuoles was found to rise markedly during the dark and to decrease during the subsequent day.[76] The molar ratio of vacuolar malic-to-isocitric acids followed a similar pattern. It is known that isocitric acid does not fluctuate in a diurnal fashion, and this acid was used as a vacuolar marker. The vacuolar location of labeled malic acid formed from $^{14}CO_2$ during the dark was also demonstrated in this study. Vacuolar accumulation of ascorbate-derived oxalate has been shown in barley.[67] L-Ascorbic acid-1-^{14}C was supplied to barley seedlings and after a period of metabolism protoplasts were isolated from the leaves. The latter were fractionated to prepare vacuoles and a fraction enriched in cytosol. About 70% of the soluble oxalic acid-^{14}C of protoplasts was recovered in vacuoles from tissues labeled for 22 and 96 hours. Oxalic acid accounted for 36 and 72%, respectively, of the ^{14}C associated with vacuoles recovered after these periods. The above-cited studies of malic acid-^{14}C and oxalic acid-^{14}C formation and subsequent accumulation in vacuoles demonstrate the feasibility of applying techniques for isolating vacuoles from protoplasts for the purpose of analyzing metabolic conversion and product compartmentation using isotopes.

Lin et al.[68] compared the ion contents of vacuoles and protoplasts of Hippeastrum and tulip petals and concluded that most of the Na^+, Mg^{++}, Ca^{++}, and Cl^- of these protoplasts is confined to the vacuole. Studies of the vacuole/cytoplasmic distribution of Na^+ and Cl^- in giant algal cells indicate that these ions are largely confined to the vacuole in these systems.[8,21] In Hippeastrum and tulip, K^+ was found at a higher level in vacuoles than in protoplasts.[68] It was suggested that this is due to K^+/H^+ exchange during vacuole isolation which leads to an equilibration of K^+ between the vacuole and the isolation medium. Equilibration would be limited by a Donnan potential (organic acids as fixed internal anions).[65] A similar phenomenon has been observed with isolated animal lysosomes.[77] It was shown[68] that the pH of petal vacuoles increased during isolation in phosphate buffer by about 3 pH units, an observation which is consistent with K^+/H^+ exchange during isolation.

All of the anthocyanin of Hippeastrum and tulip petal
protoplasts[65] and all of the betanin of red beet roots[57]
were shown to be confined to the vacuole. High concentrations of anthocyanins, flavonol glycosides, and two hydroxycinnamic acid esters were recovered in grape berry
vacuoles.[75] Grob and Matile[58] concluded that ascorbate,
glucosinolates, and phenolics of horseradish root cells are
confined to the vacuole. As discussed earlier, phenolics
were assumed to be a vacuolar marker in this tissue. The
cyanogenic glycoside dhurrin[69] and the alkaloid nicotine[33]
were shown to be vacuolar constituents in Sorghum and
tobacco, respectively. Data which have been presented to
date are consistent with the expectation that secondary
plant products are confined to the vacuole. Walker-Simmons
and Ryan[70] studied the vacuole/extravacuole distribution of
inducible protease inhibitor I in tomato and concluded that
it is localized in the vacuole, and Martinoia et al.[50] have
demonstrated the vacuolar compartmentation of nitrate in
barley.

Radiolabeled gibberellin has been recovered in vacuoles
isolated from protoplasts of prelabeled barley and cowpea
leaves by Ohlrogge et al.[56] Like the studies of Kringstad
et al.[76] and Wagner[51,67] this work demonstrates the feasibility of monitoring the subcellular fate of exogenously
administered, labeled metabolites or their metabolic products using protoplasts and vacuoles.

There has been much interest in applying vacuole/extra-
vacuole localization methods for studying enzyme compartmentation in plant cells. Much of the work to date has focused
on the subcellular location of acid hydrolases, enzymes
which were expected to be confined to the vacuole.[78] The
first study of the vacuole/extravacuole distribution of
hydrolases produced direct evidence for the vacuolar
location of acid nuclease and acid phosphatase but not for
protease or carbohydrase in Hippeastrum petals.[79] Subsequently, evidence was presented by a number of investigators
using various tissues and plants which supports the
conclusion that most of the acid hydrolase in plant cells
is confined to the vacuole. A list of enzymes localized
in vacuoles and of those found to be extravacuolar in the
systems studied is presented in Table 6. Many of the
extravacuolar enzymes listed were used to monitor vacuole
preparations for cellular contaminants. References are not

Table 6. Vacuole/extravacuole partitioning of active enzymes as determined from studies of isolated vacuoles.

Enzyme*	Percent in vacuole	Number of tissues studied	References
Acid phosphatase	30-100	8	13,33,58,60,66,79
Phosphodiesterase	55-100	4	58,60,66
Acid nuclease	50-100	3	66,79
Acid protease	80-100	4	13,15,60,64
Carboxypeptidase	most	1	66
α-Mannosidase	100	4	13,60
α-Galactosidase	80-100	2	60
β-Glucosidase	most	1	66
β-N-acetylglucosimidase	90-100	4	13,60
Acid invertase	most-100	2	57
Phytase	most	1	66
Myrosinase	29	1	58
Peroxidase	50-70	3	58,60

*Enzyme activities found to be <u>absent</u> from vacuoles in the systems tested:

Aminopeptidase[13]	Ribulose bisP carboxylase	Cytochrome c oxidase
Catalase[66]	Nitrate reductase	Glucose phosphate isomerase
Acid lipase[66]	Aldolase	Phosphoglucomutase
Alkaline lipase[66]	PEP carboxylase	Triose phosphate isomerase
Hexokinase[60]	Glc-6-PO$_4$ dehydrogenase	Succinate dehydrogenase
Glucosyl transferase[22,67]	Malate dehydrogenase	

cited in some instances. The point is again made that the
methods used for quantitation in most of these studies can
lead to significant errors. That as much as 10 or 15% of
the predominantly vacuolar hydrolase may be extravacuolar
has not been ruled out. Also, the question of how hydrolases might function in vacuoles which contain high concentrations of phenolic constituents and other inhibitory
substances has not been addressed. For example, mixing
experiments in which hydrolases are assayed in the presence
of vacuolar constituents have not been made. The turnover
of storage protein in castor bean endosperm vacuoles has
been demonstrated[10] but these structures are lacking in
secondary products such as phenolics. In certain tissues,
at least, acid phosphatase, phosphodiesterase, acid
nuclease, aminopeptidase, catalase, acid and alkaline
lipase, myrosinase, and peroxidase may not be restricted
to the vacuole.

The results listed in Tables 5 and 6 were obtained
using a variety of methods for vacuole preparation and
purification and for vacuole/extravacuole distribution
analysis. Vacuole isolation methods employed included the
use of very gentle to extreme osmotic shock and sheer to
isolate vacuoles from protoplasts and the mechanical preparation of vacuoules from plasmolyzed and fresh tissue.[16]
It is clear that isolated vacuoles prepared by various means
retain internal contents including small neutral, basic and
acidic molecules, ions, salts, proteins, sugars, a number of
glycosylated molecules, and other substances. These data
suggest that the permeability properties of the tonoplast
are unaffected by manipulation and exposure to many
different media.

Examples of natural products and substances acquired
from the environment which are accumulated by certain plants
are listed in Tables 2 and 3. Anthocyanins and the heavy
metal cadmium are representatives of these respective groups
and they provide examples of two different modes of intracellular accumulation. Anthocyanins are accumulated in the
vacuole[22,65,75] while available evidence suggests that
intracellular Cd is extravacuolar.[51] In the following discussion, these two examples are considered in some detail.

Recent studies suggest that enzymes which participate
in flavonoid biosynthesis are not associated with the sap or

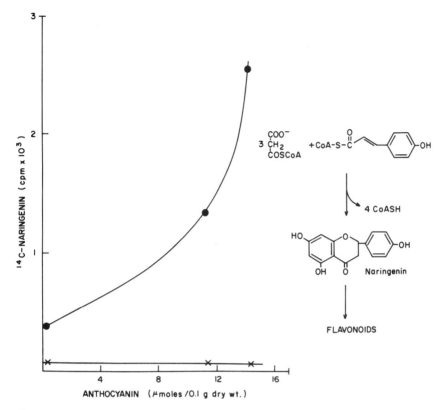

Figure 2. Relationship between anthocyanin formation and
flavanone synthase activity in subcellular fractions from
developing Hippeastrum petals. The three points of each
curve correspond to early bud, intermediate, and anthesis
stage petals containing <1, 11.4, and 14.5 μmoles antho-
cyanin/0.1 g dry wt. petal, respectively. Flavanone synthase activity of the fraction from protoplasts "enriched
in cytosol" •—• and the vacuolar sap fraction x—x are
shown. Data are from Hrazdina et al.[22]

tonoplast of isolated vacuoles.[18] [22] Figure 2 describes
results of studies by Hrazdina et al.[22] in which the
vacuole/extravacuole distribution of flavanone synthase was
studied in protoplasts of Hippeastrum petals. This enzyme
catalyzes the condensation of malonyl-CoA and p-coumaryl-

CoA to form the 15-carbon skeleton of the flavonoids. The
three points associated with each curve in Figure 2 represents fractions from early bud petals (virtually lacking in
pigment), pre-anthesis bud petals, and petals from anthesis
stage flowers (fully pigmented). The curves represent the
fraction enriched in cytosol[65] and the vacuole sap fraction.
Flavanone synthase activity increased in the fraction
enriched in cytosol recovered from protoplasts of developing
petals as the anthocyanin content of the petals increased.
No activity was observed in the vacuole sap. In other experiments (unpublished) no flavanone synthase activity was
found with tonoplast recovered from Hippeastrum or tulip
vacuoles. Similarly, no evidence has been found for the
association of UDPG:flavonol 3-O-glycosyl-transferase with
isolated tonoplast from these tissues.[18] These results do
not preclude the possibility that these enzymes are loosely
associated with tonoplast. Loosely associated components may
be lost during vacuole isolation or tonoplast manipulation.
However, no glycosyl transferase activity was found with
Hippeastrum tonoplast prepared from vacuoles produced by
osmotic shock of protoplasts in 0.2 M K_2HPO_4, pH 8, or in
0.35 M mannitol, 30 mM HEPES/NaOH, pH 8.[18] One might expect
a small amount of activity of an enzyme which is membrane
bound in vivo to be retained on that membrane after its isolation even where the membrane association is a weak one. As
noted by Marty et al.,[8] the enzyme of the flavonoid pathway,
and the synthetic pathways of several other secondary products, which is most likely to be associated with the tonoplast is the glucosyltransferase. Glycosylation is often
the last step in the formation of secondary products.[80,81]

Sumarized results of vacuole/extravacuole distribution
analysis of flavonoid biosynthetic enzymes in Hippeastrum
are presented in Table 7. The enzymes listed represent
early, intermediate, and final steps in the formation of
anthocyanin. Results for flavanone synthase, chalcone-
flavone isomerase, and UDPG: anthocyanidin glucosyltransferase were reported previously[22] and those for caffeic
acid methyltransferase and quercetin methyltransferase were
obtained in recent experiments (Hrazdina and Wagner, unpublished). Similar results have been obtained for phenylalanine ammonia-lyase but cinnamic acid 4-hydroxylase is
primarily observed in the particulate cytoplasmic fraction
of Hippeastrum protoplasts (Hrazdina and Wagner, unpublished). In these experiments with Hippeastrum a small

Table 7. Distribution of enzyme activities in subcellular fractions of Hippeastrum protoplasts.

Activity	Particulate cytoplasm	Fraction enriched cytosol	Vacuole sap	100,000 x g pellet*
		Percent of total activity recovered		
Caffeic acid methyltransferase	18	81	<1	-
Flavanone synthase	5-15	80-95	0	1-4
Chalcone-flavone isomerase	5-6	94	0	0
Quercetin methyltransferase	20	79	<1	-
UDPG: anthocyanidin glucosyltransferase	6-12	88-96	<0.1	1

*Prepared from fraction enriched in cytosol.

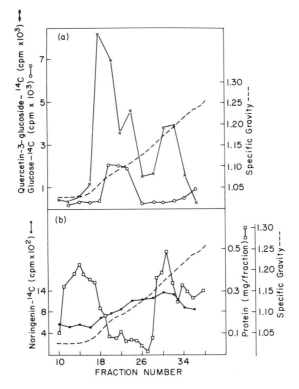

Figure 3. Glucosyltransferase, UDPG:hydrolase, flavanone
synthase, and protein associated with a particulate fraction
of a Hippeastrum petal homogenate separated on a linear
sucrose gradient. See text for details.

amount of activity was always recovered in the particulate
fraction and in material sedimented from the fraction
enriched in cytosol by centrifugation at 100,000 x g. Both
of these fractions contain membranes of the endoplasmic
reticulum[65] (unpublished observations). We have examined
endoplasmic reticulum containing fractions of Hippeastrum
tissue homogenates to determine if flavonoid biosynthetic
enzymes may be associated with this membrane. Preliminary
results are presented in Figures 3 and 4. In the experiments relating to Figure 3, anthesis stage Hippeastrum
petals (8 g fresh wt.) were homogenized with sand and 0.5 g
insoluble polyvinylpyrrolidone in ice cold 0.5 M KPO_4 buffer,

pH 8, and the homogenate was centrifuged at 5000 g for 5
min at 0°C. The supernatant was recovered and stirred with
1 g Dowex-1 (PO_4 form, pH 8) after which the 1000 g soluble
material was recovered and applied to a 15 to 60% (w/w)
sucrose gradient containing 20 mM HEPES, pH 8, 30 mM $MgCl_2$,
and 1 mM DTT. The gradient was centrifuged at 100,000 g for
16 hr at 4°C, fractionated, and assayed for UDPG: quercetin
3-O-glycosyltransferase, and UDPG: glucose hydrolase (both
in panel a of Figure 3), and for flavanone synthase and protein (panel b of Figure 3). Glucosyltransferase occurred
in the gradient in regions characteristic of smooth and
rough endoplasmic reticulum (1.11 to 1.12 and 1.15 to 1.18
g/cm^3, respectively[82]). Substantial activity was also
observed near the sample-gradient interface (1.06 to 1.07
gm/cm^3). The protein profile obtained is consistent with
the equilibration of rough endoplasmic reticulum in the
gradient at about 1.18 g/cm^3 (Figure 3b). UDPG: glucose
hydrolase was found at the sample-gradient interface and
that observed in the gradient was restricted to the region
having the density characteristic of golgi membranes (1.12
to 1.15 g/cm^3).[82] Flavanone synthase occurred throughout
the sample zone and gradient but like glucosyltransferase
it was maximal at about 1.12 and 1.18 g/cm^3. These results,
although preliminary, suggest the association of flavanone
synthase and glucosyltransferase with membranes of the
endoplasmic reticulum in Hippeastrum petals.

    The association of cinnamic acid 4-hydroxylase, an
oxidoreductase involved in flavonoid biosynthesis, with the
endoplasmic reticulum has been demonstrated in Sorghum[83]
and is suggested from the experiments of Czichi and Kindl.[84]
The latter investigators observed cooperation between
phenylalanine ammonia-lyase and cinnamic acid hydroxylase
in microsomal fractions from cucumber cotyledons. They
demonstrated that limited exchange occurs between cinnamic
acid produced from membrane-bound phenylalanine ammonia-
lyase and a soluble cinnamic acid pool suggesting "coupling"
between phyenylalanine ammonia-lyase and cinnamic acid
4-hydroxylase.[84] Possible association of cinnamic acid
4-hydroxylase with endoplasmic reticulum in Hippeastrum has
been tested and preliminary results are shown in Figure 4.
Homogenates were prepared as just described but using 0.2 M
KPO_4, pH 8 as homogenizing buffer. The supernatant after
Dowex treatment was divided into two parts. One half was
treated with 5 mM EDTA and the other with 5 mM $MgCl_2$ and

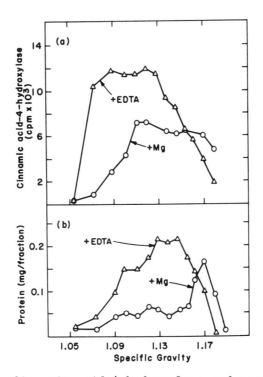

Figure 4. Cinnamic acid 4-hydroxylase and protein associated with a particulate fraction of a <u>Hippeastrum</u> petal
homogenate after its separation on a linear sucrose
gradient. One half of the homogenate was treated with 5 mM
$MgCl_2$ (panel b). The other half was treated with 5 mM EDTA
to effect a shift in the density of endoplasmic reticulum
membranes (panel a). See text for details.

membranes that were sedimented between 20,000 and 100,000 g
were recovered from each. These membrane fractions were
separated on 15 to 60% linear sucrose gradients containing
5 mM EDTA and 5 mM $MgCl_2$, respectively (both with buffer
and DTT as described above). In the absence of EDTA and
the presence of Mg most of the protein equilibrated at 1.17
g/cm³ (Figure 4b), a density typical of rough endoplasmic
reticulum.[82] Cinnamic acid 4-hydroxylase was also observed
in this region (Figure 4a) but substantial activity also
occurred at about 1.11 to 1.12 g/cm³, the density of smooth

endoplasmic reticulum.[82] In the presence of EDTA, protein
and cinnamic acid 4-hydroxylase shifted to a lower density.
Much of both equilibrated at 1.13 to 1.15 g/cm^3; however,
a large amount of cinnamic acid 4-hydroxylase moved to the
sample-gradient interface. These results suggest that
cinnamic acid 4-hydroxylase may be loosely associated with
the endoplasmic reticulum in Hippeastrum. Results similar
to those obtained for cinnamic acid 4-hydroxylase were
observed for phenylalanine ammonia lyase and flavanone
synthase (data not shown). The accumulation of both glucosyltransferase (Figure 3a) and cinnamic acid 4-hydroxylase
(Figure 4a) at the sample-gradient interface under the
conditions tested suggest that both may occur as soluble
multienzyme complexes which undergo sedimentation in the
sample zone. These observations, though preliminary,
suggest that the enzymes of the flavonoid pathway may be
loosely bound to the endoplasmic reticulum as multienzyme
complexes which are largely solubilized during vacuole/
extravacuole distribution studies such as those represented
by the data of Table 7. The results of Cutler and Conn
(this volume) suggest that cyanogenic glycosides may be
formed by multienzyme systems associated with the endoplasmic reticulum. McClure has discussed the subcellular
location of phenolic biosynthesis[84a] and Stafford has
postulated a role for multienzyme aggregates in lignin
biosynthesis.[84b]

In earlier studies of the intracellular location of
flavonoid biosynthetic enzymes, Hrazdina et al.[22] reported
poor recovery of enzyme activities from protoplast lysates
while recovery of the same activities from the "enriched
cytosol" fraction obtained during isolation of vacuoles
from these protoplasts was efficient. Marty et al.[8] have
suggested that this result implies the presence of
inhibitors in the protoplasts which may prevent the detection of enzymes present in protoplast fractions, including
the vacuole. Recent studies (Wagner and Hrazdina, unpublished) indicate that the mixing of vacuole and cytosol
fractions of Hippeastrum does not reduce the yield of
glucosyltransferase activity of the cytosol fraction. Also,
in the studies of Hrazdina et al.[22] protoplast, vacuole, and
particulate materials recovered during vacuole isolation
were exposed to polyvinylpyrrolidone and Dowex-1 prior to
assay while the fraction enriched in cytosol was not. The
last fraction was dialyzed to remove pigment and other

possible inhibitors. Dilution and subsequent concentration
and dialysis (as described in ref. 22 for the cytosol
fraction) of protoplast, cytosol and the fraction enriched
in cytosol result in the recovery of about 90% of the protoplast glucosyltransferase activity in the fraction enriched
in cytosol. No activity is recovered in the vacuole fraction
(Wagner and Hrazdina, unpublished). It is, therefore, unlikely that inhibitors prevented the expression of activity
in the vacuole fractions prepared by Hrazdina et al.[22] As
noted earlier[22] the pH optima of enzymes of the flavonoid
pathway are in the range of pH 7.5 to 8. It is highly
unlikely that these enzymes could function in the acidic
vacuolar sap in the presence of inhibitory anthocyanin.

Anthocyanins are natural products which are probably
formed in the cytoplasmic space and are subsequently transported to and accumulated in the vacuole by mechanisms which
are not understood. Transport may occur in vesicles which
originate at the site of synthesis and are transported in
the cytosol stream and finally fuse with the vacuole. Diers
et al.[85] have observed droplets of condensed tannin in
association with smooth endoplasmic reticulum in the shoot
apex of Oenothera. Similar droplets have been observed in
cell cultures of white spruce[86] and slash pine.[87]
Ginsberg[88] observed tannins in vesicles which coalesce to
form large vacuoles in root tissue of Reamuria. Recently
Pecket and Small[89] concluded that anthocyanin biosynthesis
occurs within the vacuole in vesicles known in the early
literature as cyanoplasts. These intensely pigmented structures are brightly colored in red cabbage which indicates
that their sap is very acidic. It is improbable that
flavonoid biosynthesis can occur in cyanoplasts because of
their high anthocyanin concentration and low sap pH.

Anthocyanins are an example of a natural product which
is accumulated in plants. Heavy metals are examples of
accumulated substances which are acquired from the environment. The metal cadmium is readily taken up from certain
soils by plant roots and is accumulated in all portions of
the plant.[90] Since this element is toxic to certain
critical metabolic processes, it is likely that it is
sequestered outside or within the cell. Studies of the
vacuole/extravacuole distribution of cadmium accumulated in
leaves of various plants suggest that the vacuole is not the
site of intracellular accumulation (Table 8). In these

studies plants were grown in liquid culture and pulse labeled with $^{109}CdSO_4$ and then returned to nutrient solution for 2 to 10 days. In one experiment pea plants were grown continuously in the presence of ^{109}Cd. Fractionation of leaf protoplasts indicated that intracellular cadmium was absent from vacuoles and about equally distributed between the fraction enriched in cytosol and the particulate fraction. A substantial portion of the label in leaves was found in a cell wall fraction recovered after digestion of tissue to prepare protoplasts.[51] These results suggest that Cd was primarily adsorbed to cell wall components in leaves and that portion which did enter leaf cells did not enter the vacuole. Characterization of the ^{109}Cd which occurred in the cytosol fraction of tobacco indicated that the metal was bound to 10,000 and <2000 molecular weight constituents. No evidence has been found for the existence of substantial amounts of free Cd in the cytosol in these and subsequent studies.[93] Various plants exposed to high levels of Cd during growth form large amounts of 8000 to 10,000 mol wt Cd-containing components.[51, 91-93] These components have a number of properties in common with Cd thioneins that occur in animals exposed to this metal. Cadmium thionein is an inducible Cd-binding protein which is thought to sequester Cd in the cytosol of animal cells and perhaps function in zinc homeostasis in animal tissues.[94]

The results described for Cd accumulation in plants suggest that this environmentally derived component is sequestered in cell walls and in bound form(s) in the cytosol of cells. Preliminary results suggest that zinc, which is chemically and geochemically related to Cd, may have a similar fate after uptake by plants (unpublished). In contrast, nickel may accumulate in plant vacuoles in the form of Ni citrate (Wagner, unpublished), the form of this metal found in Ni-accumulating species.[47]

MECHANISMS OF TONOPLAST TRANSPORT

The mature vacuole in most higher plants is an acidic compartment.[68, 95-97] Measures of the pH of phloem exudates which have a "cytoplasm-like" composition indicate a pH of 7 to 8.5.[95] The pH of the vacuolar sap of giant algal cells is 5.5 while that of the cytoplasm may be 7.5 (See Marty et al.[8]). Kurkdjian and Guern[96] used indicator dyes and the weak acid pH probe 5,5-dimethyloxazolidine-2,4-dione (DMO)

Table 8. Subcellular distribution of ^{109}Cd in leaf protoplasts prepared from whole plants labeled in situ.

Fraction	Percent label supplied*					
	Tomato	Pea I	Pea II	Lettuce	Tobacco	Hippeastrum
Fraction "enriched in cytosol"	0.91	0.47	0.86	0.68	2.4	2.0
Particulate	0.79	0.51	1.02	0.68	2.5	1.9
Vacuole	0	0	0	0	0.1	<0.1
Chloroplast	----	negl.	----	negl.	----	----

*Plants (except Pea II) were pulse-labeled with ^{109}CdCl$_2$ in H$_2$O and returned to Hoagland's solution for 2 to 10 days before fractionation. Pea II was continuously exposed to ^{109}CdCl$_2$ in Hoagland's for 25 days, then fractionated.

and estimated the cytoplasmic and vacuolar pH of cultured
Acer pseudoplantanus cells to be 7.6 and 3.8, respectively.

Of various mechanisms proposed for maintaining cytoplasmic pH, the most highly regarded are the active transport and the biochemical pH stat mechanisms.[95] Smith and
Raven[95] have argued that the biochemical pH stat model
(control by biochemical regulation of the number of carboxyl
groups in the cytosol) probably serves as a fine-tuning
device in maintaining cytoplasmic pH, while active proton
transport is the main means of control. Active proton
transport is thought to occur into the vacuole and out of
the cell.[95]

The mechanism of proton transport into the vacuole is
unknown but evidence supports the presence of specific
ATPase in Hippeastrum and tulip petal vacuoles,[18] [98] beet
root vacuoles,[99,100] and in membrane of lutoids of
Hevea.[101,102] In addition, ATP-dependent proton transport
into intact tulip vacuoles,[103] (manuscript submitted) and
Hevea lutoids[103a] have been reported. Certain characteristics of higher plant vacuole ATPase are presented in Table 9.
In tulip and Hippeastrum vacuoles non-specific phosphatase
is low relative to ATPase and little of the former is associated with isolated tonoplast.[18,98] Data presented in Table
9 describe the membrane-bound activity of tulip vacuoles and
activity of beet root vacuoles measured in the presence of
0.1 mM ammonium molybdate to inhibit the high level of non-
specific phosphatase present in this system.[99,100]

In both tulip and beet root, vacuolar ATPase is substrate specific and Mg dependent, and is distinguished from
non-specific phosphatase. The pH optimum is 7 to 7.5 while
that of vacuolar non-specific phosphatase in both systems
is about pH 5.5. Two major differences in the enzymes from
these systems are apparent. The tulip petal enzyme is only
10 to 30% stimulated by K^+ and is inhibited by Na^+, while
the beet root enzyme is stimulated about 100% by both ions.
The tulip enzyme is completely membrane-bound, while with
beet root equal amounts of activity are recovered in the sap
and membrane. In tulip the enzyme appears to be an integral
protein in that it is not released from the membrane by EDTA
treatment but it can be solubilized in active form using non-
ionic detergents.[98] Reduced activity of solubilized enzyme
is partially recovered on addition of phospholipids[98]

Table 9.   Characteristics of Mg-ATPase of higher plant
vacuoles.

| Characteristic | Relative ATPase | |
	Tulip petal*	Beet root**
Nucleotide specificity		
ATP	100	100
GTP	37	30
ADP	18	3
PNPP	6	3
β-Glycerol-PO$_4$	6	2
pH optimum	7.0	7.5
Ion effects		
none	0	4
Mg	100	100
Mn	41	105
Ca	2	46
Co	7	15
K	130	189
Na	64	196
Location		
membrane	100	50
sap	0	50

*Purified tonoplast measured at pH 7.0 with 3 mM substrate,
 3 mM divalent cation, and 50 mM salt (KCl except for Na
 experiment where 50 mM NaCl was used).

**Total vacuole measured at pH 8.0 in the presence of 0.1
 mM ammonium molybdate with 3 mM substrate, 3 mM divalent
 cation, and 50 mM salt (KCl excpet for Na experiment
 where 50 mM NaCl was used).

(unpublished). Further study is required to determine if
these enzymes are involved in proton or ion transport across
the tonoplast.

In addition to evidence presented supporting the existence of specific tonoplast ATPase, various investigators
have reported on ATP-stimulated solute transport in isolated
vacuoles. d'Auzac and Lioret[104] reported ATP-stimulated
DNP-sensitive citrate uptake in lutoids of Hevea. Transport
proceeded against a concentration gradient of citrate. Guy
et al.[105] reported pH-dependent ATP-stimulated 3-0-methyl-
glucose uptake in pea vacuoles. Specificity for D- over
L-glucose during transport was shown, and the proton iono-
phore SF_{6847} depressed transport. Doll et al.[106-108]
showed temperature- and pH-dependent ATP-stimulated sucrose
uptake against its concentration gradient in beet root
vacuoles. Lin et al.[103] (manuscript submitted) reported
ATP-dependent proton transport in tulip petal vacuoles.
Transport measured with intact vacuoles and tonoplast
ATPase measured in vitro have similar substrate specificity, inhibitor sensitivity and divalent ion requirements[103]
(manuscript submitted). Briskin and Leonard[109] have argued
against the existence of specific tonoplast ATPase and
suggest that tonoplast transport is mediated by permeases
which facilitate diffusion of ions and solutes down chemical
potential gradients. They do not suggest how the vacuole-
cytosol pH gradient may be maintained. These workers were
unable to demonstrate specific ATPase in tonoplast of
cultured tobacco cells.

The concepts of primary and secondary active transport
for solute transfer processes have developed from Mitchell's
hypothesis.[110] The primary active transport system converts
chemical energy - or light energy in the case of the light-
driven proton pump of Halobacterium halobium - into electro-
osmotic energy. Secondary active transport systems are
energized by a preformed gradient of a species different
from that being transported which is usually an ion gradient
and is commonly a gradient of protons. In bacteria, secondary active transport systems have been described for cations,
anions, and neutral species. Other modes of energy coupling
in solute transport include group translocation and transport
systems linked to phosphate bond energy.[110,111] Primary and
secondary active transport systems of anaerobic bacteria -
or facultative anaerobes in the absence of a terminal
electron acceptor - are shown diagrammatically at the left
of Figure 5. The organism establishes a proton motive force
(PMF) by extruding protons at the expense of ATP produced
from fermentation (Fig. 5a). In aerobes, a PMF is

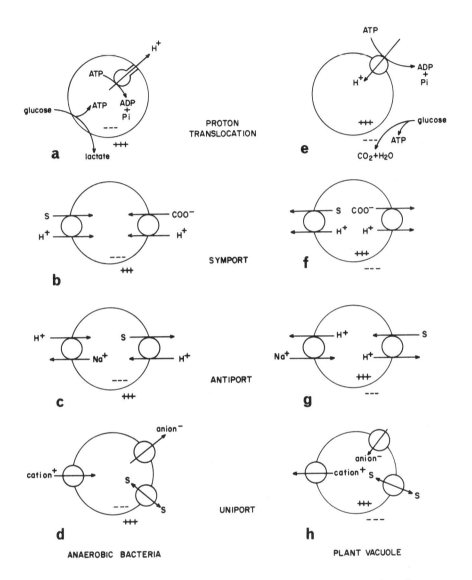

Figure 5. Models for proton motive force energized solute
transport in anaerobic bacteria (a through d) and plant
vacuoles (e through h). Adapted from reference 10.

established by extrusion of protons via electron transport,
and the ATPase complex is used to generate ATP. Secondary
active transport of solutes in aerobes and anaerobes can
occur at the expense of the PMF via symport reactions of an
electrogenic nature when uncharged solutes and protons are
co-transported or by electroneutral symport as shown for a
carboxylate and a proton (Figure 5b). Antiport transport
can occur as the electroneutral exchange of a proton for a
cation or the electrogenic exchange of a neutral solute for
a proton (Figure 5c). In a uniport reaction anions or
cations move into or out of, respectively, the cell in
response to the membrane potential and uncharged solutes
move via facilitated diffusion (Figure 5d). These
mechanisms require the participation of specific porters.
A number of these have been identified in bacterial
systems[112] and references therein.

     As a model, the plant vacuole can be compared to an
inside-out anaerobe where the cytosol of the plant cell is
analogous to the cytosol of the bacterium - right-hand side
of Figure 5. In this model a tonoplast translocating ATPase
pumps protons into the interior of the vacuole and establishes a PMF (inside positive) at the expense of ATP
produced from glycolysis and respiration (Figure 5e). The
PMF can be used to energize symports for electrogenic transport of neutral solutes to the cytoplasm or electrogenic
export of anions from the vacuole (Figure 5f). Antiports
would provide the means for electroneutral uptake of cations
such as Na^+ or K^+ and the electrogenic uptake of neutral
species (Figure 5g). Lin et al.[68] reported that the proton
gradient which exists across the tonoplast of tulip and
Hippeastrum petal vacuoles in vivo is substantially
dissipated during isolation of vacuoles in KPO_4 buffer and
that the K^+ concentration of isolated vacuoles is higher
than that of protoplasts. Wagner[65] has suggested that this
is due to an electroneutral K^+/H^+ exchange during vacuole
isolation. This process could occur via an antiport like
that diagrammed in Figure 5g. According to the model in
Figure 5h cations could be exported to the cytoplasm and
anions imported to the vacuole in response to an electrical
potential.

     In recent studies Doll and Hauer[108] deduced from
measurements of membrane potentials in isolated beet root
vacuoles that electrogenic sucrose transport in this system

is energized by ion-translocating ATPase. This conclusion
is consistent with the antiport scheme of Figure 5g.
Glucose transport in pea vacuoles may occur via an antiport
with protons as suggested by Guy et al.[105] Citrate transport in lutoids observed by d'Auzac and Lioret[104] can be
explained by an uniport mechanism (Figure 5h) in which
citrate is taken up by the vacuole in response to a membrane
potential. Both positive[95,113] and negative[108,114,115]
membrane potentials have been reported for plant vacuoles.
Alternatively, porters for organic acids may facilitate
uptake of free acids which become protonated in the vacuole
sap. Luttge and Ball[116] have concluded that the primary
system for malate accumulation in CAM vacuoles involves
active translocation of protons coupled to passive uptake of
free malate which becomes protonated in the acidic sap of
the vacuole. It is also possible that organic acids are
sequestered in vesicles (from golgi, endoplasmic reticulum
or the outer membranes of cytoplasmic organelles) at the
site of their biosynthesis. The vesicles might then fuse
with and deposit their contents into the central vacuole.
This mechanism, if it occurs, could provide the means for
accumulation of both organic acids and protons in plant
vacuoles.

The models diagrammed in Figure 5 for vacuolar transport which are analogous to those supported by evidence in
procaryotic systems can be used to explain the evidence
presented thus far for glucose and sucrose transport in
isolated plant vacuoles and citrate transport in lutoids.

An ion exchange system has been described to explain
arginine, lysine, and general anion and cation transport in
yeast vacuoles.[7,117] In this scheme polyphosphates serve
as a nondiffusable anion, and cations and anions are transported via symports and antiports with arginine. Arginine
is accumulated in yeast vacuoles and its transport is
independent of an energy source and is unaffected by the
proton ionophor dinitrophenol.[117] Polyphosphates are known
to occur in yeast vacuoles.[8] Strong evidence has been
presented recently which supports an alternative model for
arginine transport in yeast vacuoles. In this model
arginine is transported via a H^+/arginine antiport driven
by the proton motive force established by a specific
proton-translocating ATPase located in the yeast vacuolar
membrane.[117a,b]

Studies of transport in isolated vacuoles are in a preliminary stage. The experience gained in studies of transport in procaryotic systems and the availability of isolated vacuoles from a number of types of plant tissues should facilitate rapid growth in our understanding of higher plant tonoplast transport. Vacuolar accumulation of complex molecules such as secondary plant products and proteins may proceed via complex means. The former may reach the tonoplast packaged in vesicles which fuse with the vacuole and the latter may be introduced by the mechanism described by the signal hypothesis.[118,119]

CONCLUDING REMARKS

The mature plant cell vacuole has many functions, not the least of which are its roles in metabolite storage and sequestration. Methods for isolating vacuoles have been developed in recent years which permit characterization of vacuole sap and tonoplast components, vacuole/extravacuole distribution analysis, and studies of tonoplast transport. Although difficulties exist in applying these methods and in performing these analyses, much progress has been made in just the last few years in characterizing the nature of vacuolar constituents and enzymes. Direct studies of tonoplast transport are beginning to elucidate the mechanisms underlying the storage and sequestration functions of this important plant cell compartment.

REFERENCES

1. Zirkle, C. 1937. The Plant Vacuole. Bot. Rev. 3: 1-30.
2. Guilliermond, A. 1941. The Cytoplasm of the Plant Cell. Chronica Botanica Co., Waltham, MA. pp. 1-247.
3. Kramer, P. J. 1955. Physical chemistry of the vacuoles. In Encyclopedia of Plant Physiology (W. Ruhland, ed.). Springer-Verlag, Berlin, Vol 1, pp. 649-660.
4. Voeller, B. R. 1964. The plant cell: aspects of its form and function. In The Cell (J. Brachet, A. E. Mirsky, eds.) Academic Press, New York, Vol. VI, pp. 245-312.

5. De Robertis, E. D. P., W. W. Narinski, F. A. Salz.
    1965. The plant vacuole. In Cell Biology, 4th Ed.
    W. B. Saunders Co., Philadelphia, PA.
6. Pisek, A. 1955. Chemie des Zellsaftes. In Encyclo-
    pedia of Plant Physiology (W. Ruhland, ed.) Vol. I,
    Springer-Verlag, Berlin, pp. 614-626.
7. Matile, P. 1978. Biochemistry and function of
    vacuoles. Ann. Rev. Plant Physiol. 29: 193-213.
8. Marty, F., D. Branton, R. A. Leigh. 1980. Plant
    vacuoles. In The Biochemistry of Plants (P. K.
    Stumpf, E. E. Conn, eds.). Vol. I, Academic Press,
    New York, pp. 625-658.
9. Leigh, R. A. 1979. Do plant vacuoles degrade cyto-
    plasmic components? Trends in Biol. Sci. N37-N38.
10. Nishimura, M., H. Beevers. 1979. Hydrolysis of protein
    in vacuoles isolated from higher plant tissue.
    Nature 277: 412-413.
11. Van der Wilden, W., E. M. Herman, M. J. Chrispeels. 1980.
    Protein bodies of mung bean cotyledons as autophagic
    organelles. Proc. Nat. Acad. Sci. USA 77: 428-432.
12. Peoples, M. B., V. C. Beilharz, S. P. Waters, R. J.
    Simpson, M. J. Dalling. 1980. Nitrogen redistri-
    bution during grain growth in wheat. Planta
    149: 241-251.
13. Heck, V., E. Martinoia, P. Matile. 1981. Subcellular
    localization of acid protease in barley mesophyll
    protoplasts. Planta 151: 198-200.
14. Ragster, L. E., M. J. Chrispeels. 1981. Autodigestion
    in crude extracts of soybean leaves and isolated
    chloroplasts as a measure of proteolytic activity.
    Plant Physiol. 67: 104-109.
15. Lin, W., V. A. Wittenbach. 1981. Subcellular locali-
    zation of proteases in wheat corn mesophyll proto-
    plasts. Plant Physiol. 67: 969-972.
16. Wagner, G. Isolation of higher plant vacuoles and
    tonoplast. In Isolation of Membranes and Organelles
    from Plant Cells (J. L. Hall, A. L. Moore, eds.).
    Academic Press, London (In press).
17. Wagner, G. J., H. C. Butcher, H. W. Siegelman. 1978.
    The plant protoplast. BioScience 28: 95-101.
18. Wagner, G. J. 1981. Enzymic and protein character of
    tonoplast from Hippeastrum vacuoles. Plant Physiol.
    68: 499-503.
19. Wiebe, H. H. 1978. The significance of plant vacuoles.
    BioScience 28: 327-331.

20. Walter, H., E. Stadelmann. 1968. The physiological prerequisites for the transition of autotrophic plants from water to terrestrial life. BioScience 18: 694-701.

21. Dainty, J. 1968. The structure and possible function of the vacuole. In Plant Cell Organelles (J. B. Pridham, ed.). Academic Press, New York, pp. 40-46.

22. Hrazdina, G., G. J. Wagner, H. W. Siegelman. 1978. Subcellular localization of enzymes of anthocyanin biosynthesis in protoplasts. Phytochemistry 17: 53-56.

23. Schmeltz, I. 1971. Nicotine and other tobacco alkaloids. In Naturally Occurring Insecticides (J. Jacobson, D. G. Crosby, eds.). Dekker, New York, pp. 559-585.

24. Muller, C. H. 1969. The "co" in coevolution. Science 164: 197-198.

25. Fraenkel, G. S. 1959. The raison d'etre of secondary plant substances. Science 129: 1466-1470.

26. Rhoades, D. F., R. G. Cates. 1976. Toward a general theory of plant antiherbivore chemistry. In Biochemical Interaction between Plants and Insects (J. W. Wallace, R. L. Mansell, eds.). Recent Advances in Phytochemistry, Vol. 10. Plenum Press, New York, pp. 168-213.

27. Bell, A. O. 1981. Biochemical mechanisms of disease resistance. Annu. Rev. Plant Physiol. 32: 21-81.

28. Berenbaum, M., P. Feeny. 1981. Toxicity of angular furanocoumarins to swallowtail butterflies: escalation in a coevolutionary arms race? Science 212: 927-929.

29. Franceschi, V. R., H. T. Horner. 1980. Calcium oxalate crystals in plants. Bot. Rev. 46: 361-427.

30. Oaks, A., R. G. S. Bidwell. 1970. Compartmentation of intermediary metabolites. Annu. Rev. Plant Physiol. 21: 43-66.

31. Osmond, G. B. 1976. Transport in plants. In Encyclopedia of Plant Physiology, New Series. U. Luttge, M. G. Pitman, eds. Vol. 2A Springer-Verlag, Berlin, pp. 347-372.

32. Kisaki, T., N. E. Tolbert. 1969. Glycolate and glyoxylate metabolism by isolated peroxisomes and chloroplasts. Plant Physiol. 44: 242-250.

33. Saunders, J. A. 1979. Investigations of vacuoles isolated from tobacco. Plant Physiol. 64: 74-78.

34. Saunders, J. A., E. E. Conn, C. H. Lin, C. R. Stocking.
    1977. Subcellular localization of the cyanogenic
    glucoside of Sorghum by autoradiography. Plant
    Physiol. 59: 647-652.
35. Robinson, T. 1980. The Organic Constituents of Higher
    Plants. Edit. 4. Cordus Press, P. O. Box 587,
    North Amherst, MA, pp. 201-205.
36. Imai, K., K. Furuya. 1951. Study of the phytochemical
    component of Fagopyrum cymosum Meisn. J. Pharm.
    Soc. Japan 71: 266-273.
37. Asen, S., R. N. Stewart, K. H. Norris. 1977. Antho-
    cyanin and pH involved in the color of 'heavenly
    blue' morning glory. Phytochemistry 16: 1118-1119.
38. Burch, G. E. 1972. Experiments of nature: whole
    leaf and purified alkaloids. Am. Heart J.
    83: 845-847.
39. Akahori, A., F. Yasuda, M. Togami, K. Kagawa, T.
    Okahishi. 1969. Variation in isodiotigenin and
    diosgenin content in aerial parts of Dioscorea
    tokoro. Phytochemistry 8: 2213-2217.
40. Feeny, P. P. 1968. Seasonal changes in the tannin
    content of oak leaves. Phytochemistry 7: 871-880.
41. Arnold, G. W., J. J. Hill. 1972. Chemical factors
    effecting selection of food plants by ruminants.
    In Phytochemical Ecology (J. Harborne, ed.).
    Academic Press, London and New York, pp. 71-101.
42. Fassett, D. W. 1973. Oxalates. In Toxicants Occur-
    ring Naturally in Foods, National Academcy of
    Sciences, 2101 Constitution Ave., N.W.,
    Washington, D.C., pp. 346-362.
43. Liener, I. E., J. E. Rose. 1953. Soyin, a toxic
    protein from the soybean III. Immunochemical pro-
    perties. Proc. Soc. Exp. Biol. Med. 83: 539-547.
44. Vinson, C. G., F. B. Cross. 1942. Vitamin C content
    of persimmon leaves and fruits. Science 96: 430-431.
45. Fassett, D. W. 1973. Nitrates and Nitrites. In
    Toxicants Occurring Naturally in Foods, National
    Academy of Sciences, 2101 Constitution Ave., N. W.,
    Washington, D.C., pp. 7-25.
46. Allaway, W. H., H. A. Laitinen, H. W. Lakin, O. H. Muth.
    1974. Selenium. In Geochemistry and The Environ-
    ment, Vol. 1. National Academy of Sciences, 2101
    Constitution Ave., Washington, D.C., pp. 57-63.

47. Brooks, R. R., J. Lee, R. D. Reeves. 1976. *Sebertia acuminata*: a hyperaccumulator of nickel from New Caledonia. Science 193: 579-580.
48. Page, A. L., F. T. Bingham, C. Nelson. 1972. Cadmium absorption and growth of various plant species as influenced by solution cadmium concentration. J. Environ. Quality 1: 288-291.
49. Finlayson, D. G., H. R. MacCarthy. 1973. Pesticide residues in plants. In Environmental Pollution by Pesticides (C. A. Edwards, ed.). Plenum Press, New York, pp. 57-86.
50. Martinoia, E., U. Heck, A. Wiemken. 1981. Vacuoles as storage compartments for nitrate in barley leaves. Nature 289: 292-294.
51. Wagner, G. J. 1979. The subcellular site and nature for intracellular cadmium in plants. In Trace Substances in Environmental Health - XIII (D. D. Hemphill, ed.). Univ. of Missouri Press, Columbia, Missouri, pp. 115-123.
52. Wagner, G. J., H. W. Siegelman. 1975. Large-scale isolation of intact vacuoles and isolation of choloroplasts from mature plant tissues. Science 190: 1298-1299.
53. Leigh, R. A., D. Branton. 1976. Isolation of vacuoles from root storage tissue of *Beta vulgaris* L. Plant Physiol. 58: 656-662.
54. Lorz, H. C., T. Harms, I. Potrykus. 1976. Isolation of "Vacuoplasts" from protoplasts of higher plants. Biochem. Physiol. Pflanzen 169: 617-620.
55. Leigh, R. A., D. Branton, F. Marty. 1979. Methods of isolation of intact vacuoles and fragments of tonoplast. In Plant Organelles, Methodological Surveys (B) Biochemistry, Vol. 9 (E. Reid, ed.). Ellis Harwood LTD., Publishers, Chichester, West Sussex, England, pp. 69-80.
56. Ohlrogge, J. B., J. L. Garcia-Martinez, D. Adams, L. Rappaport. 1980. Uptake and subcellular compartmentation of gibberellin A_1 applied to leaves of barley and cowpea. Plant Physiol. 66: 422-427.
57. Leigh, R. A., T. Rees, W. A. Fuller, J. Banfield. 1979. The location of acid invertase activity and sucrose in the vacuoles of storage of roots of beet root (*Beta vulgaris*). Biochem. J. 178: 539-547.

58. Grob, K., P. Matile. 1980. Compartmentation of
    ascorbic acid in vacuoles of horseradish root cells.
    Z. Pflanzenphysiol. 98: 235-243.
59. Buser, C., P. Matile. 1977. Malic acid in vacuoles
    isolated from Bryophyllum leaf cells. Z.
    Pflanzenphysiol. 82: 462-466.
60. Boller, T., H. Kende. 1979. Hydrolytic enzymes in
    the central vacuole of plant cells. Plant Physiol.
    63: 1123-1132.
61. Taylor, A. R. D., J. L. Hall. 1976. Some physio-
    logical properties of protoplasts isolated from maize
    and tobacco tissues. J. Exp. Bot. 27: 383-391.
62. Hall, J. L. 1979. Methods for isolation of proto-
    plasts and plasma membranes. In Plant Organelles,
    Methodological Surveys (B) Biochemistry, Vol. 9 (E.
    Reid, ed.). Ellis Horwood Ltd. Publisher,
    Chichester, West Sussex, England. pp. 69-80.
63. Cocking E. C. Isolation of Plant Protoplasts. In
    Isolation of Membranes and Organelles from Plant
    Cells (J. L. Hall, A. L. Moore, eds.). Academic
    Press, London (In Press).
64. Wagner, G. J., P. Mulready, J. Cutt. 1981. Vacuole/
    extravacuole distribution of soluble protease in
    Hippeastrum petal and Triticum leaf protoplasts.
    Plant Physiol. 68: 1081-1089.
65. Wagner, G. J. 1979. Content and vacuole/extravacuole
    distribution of neutral sugars, free amino acids
    and anthocyanin in protoplasts. Plant Physiol.
    64: 88-93.
66. Nishimura, M., H. Beevers. 1978. Hydrolases in
    vacuoles from castor bean endosperm. Plant Physiol.
    62: 44-48.
67. Wagner, G. J. 1981. Vacuolar deposition of ascorbate-
    derived oxalic acid in barley. Plant Physiol.
    67: 591-593.
68. Lin, W., G. J. Wagner, H. W. Siegelman, G. Hind. 1977.
    Membrane-bound ATPase of intact vacuoles and tono-
    plasts isolated from mature plant tissue. Biochim.
    Biophys. Acta 465: 110-117.
69. Saunders, J. A., E. E. Conn. 1978. Presence of the
    cyanogenic glucoside dhurrin in isolated vacuoles
    from Sorghum. Plant Physiol. 61: 154-157.
69a. Galun, E. 1981. Plant protoplasts as physiological
    tools. Annu. Rev. Plant Physiol. 32: 237-266.

70. Walker-Simmons, M., C. A. Ryan. 1977. Immunological identification of protease inhibitors I and II in isolated tomato leaf vacuoles. Plant Physiol. 60: 61-63.
71. Beevers, L. 1976. Nitrogen Metabolism in Plants. Arnold Press, London.
72. Holleman, J. M., J. L. Key. 1967. Inactive and protein precursor pools of amino acids in the soybean hypocotyl. Plant Physiol. 42: 29-36.
73. Giaquinta, R. 1978. Source of sink leaf metabolism in relation to phloem translocation. Plant Physiol. 61: 380-385.
74. Outlaw, W. H., D. B. Fisher, A. L. Christy. 1975. Compartmentation in Vica faba leaves. Plant Physiol. 55: 704-711.
75. Moskowitz, A. H., G. Hrazdina. 1981. Vacuolar contents of fruit subepidermal cells from Vitis species. Plant Physiol. 68: 686-692.
76. Kenyon, W. H., R. Kringstad, C. C. Black. 1978. Diurnal changes in the malic acid content of vacuoles isolated from leaves of the crassulacean acid metabolism plant, Sedum telephium. FEBS Lett. 94: 281-283.
77. Reijngoud, D. J., J. M. Tager. 1977. The permeability properties of the lysosomal membrane. Biochim. Biophys. Acta 472: 419-449.
78. Matile, P. 1975. The Lytic Compartment of Plant Cells. Springer-Verlag, New York, pp. 1-175.
79. Butcher, H. C., G. J. Wagner, H. W. Siegelman. 1977. Localization of acid hydrolases in protoplasts. Plant Physiol. 59: 1098-1103.
80. Swain, T. 1976. Nature and properties of flavonoids. In Chemistry and Biochemistry of Plant Pigments (T. W. Goodwin, ed.). Edit. 2, Vol. 1, Academic Press, New York, pp. 425-463.
81. Conn, E. E. 1973. Biosynthesis of cyanogenic glucosides. Biochem. Soc. Symp. 38: 277-302.
82. Quail, P. H. 1979. Plant cell fractionation. Annu. Rev. Plant Physiol. 30: 425-484.
83. Saunders, J. A., E. E. Conn, C. H. Lin, M. Shimada. 1977. Localization of cinnamic acid 4-monoxygenase and the membrane-bound enzyme system for dhurrin biosynthesis in Sorghum seedlings. Plant Physiol. 60: 629-634.

84. Czichi, V., H. Kindl. 1977. Phenylalanine ammonia
     lyase and cinnamic acid hydroxylases as assembled
     consecutive enzymes on microsomal membranes of
     cucumber cotyledons: cooperation and subcellular
     distribution. Planta 134: 133-143.
84a. McClure, J. W. 1977. The physiology of phenolic com-
     pounds in plants. Recent Advan. Phytochem.
     12: 525-556.
84b. Stafford, H. A. 1974. Possible multienzyme complexes
     regulating the formation of C_6-C_3 phenolic compounds
     and lignins in higher plants. Recent Advan.
     Phytochem. 8: 53-79.
85. Diers, L., F. Schotz, B. Meyer. 1973. Uber die
     Ausbildung von Gerbsstoffvakuolen bei Oenothera.
     Cytobiologie 7: 10-19.
86. Chafe, S. C., D. J. Durzan. 1973. Tannin inclusions
     in cell suspension cultures of white spruce. Planta
     113: 251-262.
87. Baur, P. S., C. H. Walkinshaw. 1974. Fine structure
     of tannin accumulation in callus cultures of Pinus
     elliotti (slash pine). Can. J. Bot. 52: 615-619.
88. Ginsberg, C. 1967. The relation of tannins to the
     differentiation of the root tissues in Reaumuria
     palastina. Bot. Gaz. 128: 1-10.
89. Pecket, R. C., C. J. Small. 1980. Occurrence,
     localization and development of anthocyanoplasts.
     Phytochemistry 19: 2571-2576.
90. Haghiri, F. 1973. Cadmium uptake by plants. J.
     Environ. Qual. 2: 93-96.
91. Bartlof, M., E. Brennan, C. A. Price. 1980. Partial
     characterization of cadmium-binding protein from
     the roots of cadmium-treated tomato. Plant Physiol.
     66: 438-441.
92. Weigel, H. J., H. J. Jager. 1980. Subcellular distri-
     bution and chemical form of cadmium in bean plants.
     Plant Physiol. 65: 480-482.
93. Wagner, G. J., M. M. Trotter. Inducible cadmium
     binding complexes of cabbage and tobacco. Plant
     Physiol. (In Press).
94. Cherian, M. G. 1979. Metabolism and potential toxic
     effects of metallothionein. In Metallothionein
     (J. H. R. Kagi, M. Nordberg, eds.). Birkhauser
     Verlag, Basel, Boston, Stuttgart. pp. 337-345.

95. Smith, F. A., J. A. Raven. 1979. Intracellular pH
    and its regulation. Annu. Rev. Plant Physiol.
    30: 289-311.
96. Kurkdjian, A., J. Guern. 1978. Intracellular pH in
    higher plant cells. Plant Sci. Lett. 11: 337-344.
97. Drawert, H. 1955. Der pH-Wert des Zellsaftes. In
    Encyclopedia of Plant Physiology (W. Ruhland, ed.).
    Springer-Verlag, Berlin, Vol. 1, pp. 627-639.
98. Wagner, G. J., P. Mulready. 1981. Solubilization and
    characterization of tonoplasts ATPase. Plant Physiol.
    Supp. 67: 8.
99. Leigh, R. A., R. R. Walker. 1980. ATPase and acid
    phosphatase activities associated with vacuoles
    isolated from storage roots of red beet (Beta
    vulgaris L.). Planta 150: 222-229.
100. Walker, R. R., R. A. Leigh. 1981. Characterization
    of a salt-stimulated ATPase activity associated
    with vacuoles isolated from storage roots of red
    beet (Beta vulgaris L.). Planta (In Press).
101. d'Auzac, J. 1975. Characterisation d'une ATPase
    membranaire en presence d'une phosphatase acide
    dans les lutoides du latex d'Hevea brasiliensis.
    Phytochemistry 14: 671-675.
102. d'Auzac, J. 1977. ATPase membranaire de vacuoles
    lysosomales: les lutoides du latex d'Hevea
    brasiliensis. Phytochemistry 16: 1881-1885.
103. Lin, W., G. J. Wagner, G. Hind. 1977. The proton
    pump and membrane potential of vacuoles isolated
    from Tulipa petals. Plant Physiol. Supp. 59: 85.
103a. Marin, B., M. Marin-Lanza, E. Komor. 1981. The
    proton motive potential difference across the vacuo-
    lysosomal membrane of Hevea brasiliensis (rubber tree)
    and its modification by a membrane-bound adenosine
    triphosphatase. Biochem. J. 98: 365-372.
104. d'Auzac, J., C. Lioret. 1974. Mise en evidence d'un
    mecanisme d'accumulation du citrate dans les
    lutoides du latex d'Hevea brasiliensis. Physiol.
    Veg. 12: 617-635.
105. Guy, M., L. Reinhold, D. Michaeli. 1979. Direct evi-
    dence for a sugar transport mechanism in isolated
    vacuoles. Plant Physiol. 64: 61-64.
106. Doll, S., F. Rodier, J. Willenbrink. 1979. Accumulation
    of sucrose in vacuoles isolated from red beet
    tissue. Planta 144: 407-411.

107. Willenbrink, J., S. Doll. 1979. Characteristics of
     the sucrose uptake system of vacuoles isolated
     from red beet tissue. Planta 147: 159-162.
108. Doll, S., R. Hauer. 1981. Determination of the mem-
     brane potential of vacuoles isolated from red beet
     storage tissue. Planta 152: 153-158.
109. Briskin, D. P., R. T. Leonard. 1980. Isolation of
     tonoplast vesicles from tobacco protoplasts. Plant
     Physiol. 66: 684-687.
110. Rosen, B. P., E. R. Kashket. 1978. Energetics of
     active transport. In Bacterial Transport (B. P.
     Rosen, ed.). Chap. 12, Marcel Dekker, Inc.,
     New York. pp. 559-620.
111. Hays, J. B. 1978. Group translocation transport
     systems. In Bacterial Transport (B. P. Rosen, ed.).
     Chap. 2, Marcel Dekker, Inc., New York. pp. 43-102.
112. Goto, K., H. Kirata, Y. Kagawa. 1980. A stable Na^+/H^+
     antiporter of thermophylic bacterium PS3. J.
     Bioenerget. Biomembran. 12: 297-308.
113. Mertz, S. M., N. Higinbotham. 1976. Transmembrane
     electropotential in barley roots as related to cell
     type, cell location, and cutting and aging effects.
     Plant Physiol. 57: 123-128.
114. Goldsmith, M. H. M., R. E. Clealand. 1978. The
     contribution of tonoplast and plasmamembrane to
     the electrical properties of higher plant cells.
     Planta 143: 261-265.
115. Saftner, R. A., R. E. Wyse. 1980. Alkali cation/
     sucrose cotransport in the root sink of sugar beet.
     Plant Physiol. 66: 884-889.
116. Luttge, U., E. Ball. 1979. Electrochemical investi-
     gation of active malic acid transport at the tono-
     plast into the vacuoles of the CAM plant Kalanchoe
     daigremontiana. J. Membrane Biol. 47: 401-422.
117. Boller, T., M. Durr, A. Wiemken. 1975. Character-
     ization of a specific transport system for arginine
     in isolated yeast vacuoles. Eur. J. Biochem.
     54: 81-91.
117a. Ohsumi, Y., Y. Anraku. 1981. Active transport of
     basic amino acids driven by a proton motive force
     in vacuolar membrane vesciles of Saccharomyces
     cerevisiae. J. Biol. Chem. 256: 2079-2082.

117b. Kakinuma, Y., Y. Ohsumi, Y. Anraku. 1981. Properties of H^+-translocating adenosine triphosphatase in vacuolar membranes of Saccharomyces cerevisiae. J. Biol. Chem. 256: 10859-10863.

118. Blobel, G., B. Dobberstein. 1975. Transfer of proteins across membranes. J. Cell. Biol. 67:852-862.

119. Blobel, G. 1980. Intracellular protein topogenesis. Proc. Natl. Acad. Sci. USA 77: 1496-1500.

Chapter Two

THE NATURE OF THE CYANIDE-RESISTANT PATHWAY IN PLANT
MITOCHONDRIA

JAMES N. SIEDOW

Department of Botany
Duke University
Durham, NC   27706

INTRODUCTION

Plant mitochondria have long been recognized as being
similar to their more extensively studied counter-parts in
animals with respect to both form and function. However
differences between plant and animal mitochondria have been
reported. At one time, these differences were not felt to
be real, but rather to be artifacts due to difficulties
associated with isolating mitochondria from plant tissue.
This view is no longer widely held, and as a result, it is
now recognized that there are a number of distinct differences between plant and animal mitochondria. In this
review, I will first outline several features which are
specifically associated with the plant mitochondrial electron transfer chain. I will then take one of these features, the cyanide-resistant path of electron flow to
oxygen, and discuss in more detail what is currently known
regarding the nature of the pathway and the unusual cyanide-
resistant oxidase which is apparently associated with it.

PLANT MITOCHONDRIAL ELECTRON TRANSFER

The plant mitochondrial electron transfer chain appears
similar to that found in mitochondria from more extensively
studied animal sources such as beef heart or rat liver. It
contains four multicomponent electron transfer complexes
(Figure 1) all tightly associated with the inner mitochon-
diral membrane.[1,2] This chain allows the transfer of
reducing equivalents from TCA cycle substrates to molecular
oxygen. Substrates such as malate or α-ketoglutarate are
referred to as NAD-linked because their oxidations are
coupled to the reduction of NAD to NADH which is subsequently oxidized through the NADH dehydrogenase associated
with complex I. Succinate on the other hand is oxidized
directly by complex II. Both NADH and succinate are formed
in the mitochondrial matrix so their initial oxidation by
the electron transfer chain takes place on the inner surface
of the inner membrane (Figure 2). Note that the linear
sequence of electron carriers implied in Figure 1 is an
oversimplification. Electron flow, particularly in the
ubiquinone region between complexes I and II and complex
III, seems not to involve a strict linear chain.[3] As with
animal mitochondria, three sites of energy conservation are
found in plant mitochondria.[4] Each "site" represents the
point of generation of a transmembrane proton gradient and
these are located at complexes I, III and IV. The proton

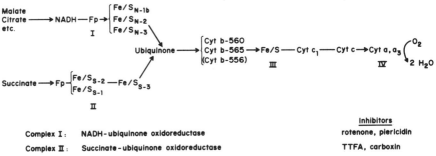

Figure 1. Sequence of electron carriers on the cyanide-
sensitive electron transfer pathway in plant mitochondria.
Fe/S = iron-sulfur protein, Fp = flavoprotein.

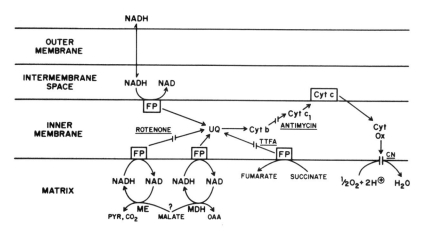

Figure 2.   Schematic representation of the topographic orientation of the dehydrogenases on the inner mitochondrial membrane of higher plants.   Adapted from Palmer.[1]

gradient is subsequently dissipated through another inner membrane-bound protein complex (F_1F_0) which couples the breakdown of the gradient to the synthesis of ATP in the mitochondrial matrix.[5]   Inhibitors commonly utilized to block electron transfer in each of the four complexes are listed in Figure 1.

While these basic similarities are well documented (see refs. 1 and 2 for recent reviews), a number of distinct differences between plant and animal mitochondria appear on closer inspection, and it is these differences which have, in no small way, accounted for the effort expended by plant "mitochondriacs" in recent years.   Among these differences are some dissimilarities in the composition of the electron transfer chains.   Considering the cytochromes, there appear to be 3 b-type cytochromes in plant mitochondria (cytochromes b-556, b-560 and b-565; room temperature maxima)[2,6] while only 2 appear in animal mitochondria.[7]   Since the mechanism by which the 2 b-type cytochromes act in animal mitochondria is far from understood, the role of the additional cytochrome in plant mitochondria is, needless to say, unknown.   On the basis of potentiometric titrations[8,9] and general behavior upon addition of antimycin A,[9] it is felt that cytochrome b-560 ($E_{m7.2}$ = +42 to 79 mV) and cytochrome

$\underline{b}$-565 ($E_{m7.2}$ = -75 mV) correspond to the high and low
potential cytochromes $\underline{b}$-561 and $\underline{b}$-566 of mammalian mitochondria, leaving cytochrome $\underline{b}$-556 as the "unusual" one.
The midpoint potential of cytochrome $\underline{b}$-556 is too high
($E_{m7.2}$ = +75 to 80 mV) to attribute it to the low wavelength
component of the split α-band observed in cytochrome $\underline{b}$-566
at liquid nitrogen temperatures.[10] However, this latter
explanation probably accounts for the additional minor
b-cytochrome species (cyt $\underline{b}$-558) reported by Lambowitz and
Bonner[9] in mung bean (<u>Vigna</u> <u>radiata</u>, formerly and probably
more widely recognized as <u>Phaseolus</u> <u>aureus</u>) and white potato
(<u>Solanum</u> <u>tuberosum</u>) mitochondria. To add to the confusion,
a fourth b-type cytochrome, which is reduced only upon
addition of a low potential reductant such as dithionite,
has been reported in plant mitochondria isolated from a
variety of sources.[9,11] A detailed description of the
redox and kinetic properties of all the plant mitochondrial
cytochromes has recently appeared.[2]

The current understanding of the flavoprotein complement in plant mitochondria could charitably be described as
"a mess", and is not as well defined as in animal mitochondria. Suffice it to say, that differences between plant
and animal mitochondrial flavoproteins do exist and that
Storey[2] has attempted to clarify the situation, as best it
can be, in his recent review.

With respect to iron-sulfur centers, the picture is
currently quite clear. Extensive studies combining potentiometric titrations with EPR spectroscopy by Cammack and
Palmer[12,13] and later by Rich and Bonner[14] led to a
reasonably good understanding of the iron-sulfur centers
present in plant mitochondria. All three of the iron-sulfur
centers found in complex II in mammalian mitochondria (S-1,
S-2, and S-3[15]) are present in higher plant mitochondria
while in complex I, apparently only the very low potential
center N-1a is absent in plants.[14] While plant complexes I
and II seemed to show the same iron-sulfur proteins found
in animal mitochondria, there was no evidence that plant
mitochondria contained the EPR signal associated with the
lone iron-sulfur protein found in complex III in animal
mitochondria,[14] the "Rieske" iron-sulfur center.[16] This
apparent absence became more surprising in light of later
evidence supporting the central role of the Rieske center
mediating electron flow between ubiquinone and cytochrome

c_1 in animal mitochondria. This discrepancy was cleared up recently by Prince et al.[18] who were able to generate the EPR signal associated with the "Rieske" iron-sulfur center upon addition of 5-n-undecyl-6-hydroxy-4,7-dioxobenzothiazole (UHDBT) to reduced mung bean or potato mitochondria. UHDBT was also reported to be a potent inhibitor of plant mitochondrial electron transfer. UHDBT had previously been found to inhibit electron transfer in yeast[19] and heart[20] mitochondria and in purple photosynthetic bacteria.[21] In the photosynthetic bacteria, UHDBT was found to act at the level of the Rieske-type iron-sulfur protein present in that system possibly binding directly to the center.[22] In heart mitochondria, UHDBT acts at complex III blocking electron flow to cytochrome c[20] and preventing the "oxidant-induced reduction" of the b-type cytochromes.[23] Both of the latter observations are compatible with UHDBT inhibiting at the level of the Rieske center in these mitochondria.

In addition to oxidizing matrix-derived (internal) NADH, most plant mitochondria are able to directly oxidize externally added NADH. This is a function not found in intact animal mitochondria and is due to the presence of an NADH dehydrogenase bound to the outer surface of the inner mitochondrial membrane.[1,24] Electrons from external NADH feed into the main electron transfer chain at the level of ubiquinone and are then transferred via complexes III and IV to oxygen (Figure 2). The first of the three sites of energy conservation (i.e., complex I) is bypassed during the oxidation of external NADH resulting in less ATP produced per molecule of oxygen reduced relative to the oxidation of internal NADH. Little is known about the nature of this dehydrogenase, how it interacts with the main electron transfer chain, and what sort of controls might operate on it in vivo.[1,24] Earlier studies noted a stimulatory effect of cations, particularly Ca^{++}, on exogenous NADH oxidation[25,26] which led to speculation regarding its regulation in vivo.[1] Recent evidence suggests that the stimulatory effect of Ca^{++} (and other cations) is due to the more general screening of fixed negative charges on the membrane surface, which would otherwise repel the approach of the anionic NADH.[27]

Electron transfer through complex I in animal mitochondria is markedly inhibited by either rotenone or piericidin A (Figure 1) while most plant mitochondria show

only a partial inhibition of internal NADH oxidation upon
addition of either inhibitor. This seems to result from the
presence of two separate NADH dehydrogenases on the inner
surface of the inner mitochondrial membrane; only one of
which is sensitive to rotenone or piericidin A.[1,24] Further,
the rotenone-resistant pathway seems to bypass the first
site of energy conservation.[1] Why two internal NADH
dehydrogenases exist in plants isn't fully known, but it is
possibly associated with the presence of two enzymes for the
oxidation of malate in the mitochondrial matrix, malate
dehydrogenase (MDH) and an NAD-dependent malic enzyme.[28]
There is also evidence that each of the latter enzymes is to
some extent associated with one specific NADH dehydrogenase,
but the issue is a bit confusing. Brunton and Palmer[28]
believe NADH derived via malic enzyme activity is linked to
the rotenone-sensitive NADH dehydrogenase while MDH-derived
NADH is preferentially oxidized by the rotenone-resistant
dehydrogenase. Recent studies of Rustin et al.[29] seem to
indicate just the opposite. This is still a very active
area of research and the reader should consult the two
reviews of Palmer[1,24] and the recent literature for more
detailed discussions. Figure 2 presents a diagrammatic
representation of how the different dehydrogenases are
currently felt to be organized on the inner mitochondrial
membrane in plants.

An additional aspect in which plant mitochondria differ
from most animal mitochondria and for which plant mitochondria are probably best known, is the presence of a cyanide-
resistant, "alternative" electron transfer pathway. Cyanide-
resistant respiration has been the subject of a number of
reviews in recent years starting with the thorough treatment
of Henry and Nyns in 1975,[30] followed by Solomos in 1977[31]
and both Day et al.[32] and Lambers[33] in 1980. Other reviews,
which dealt more generally with plant mitochondria have also
considered various aspects of the cyanide-resistant
pathway.[1,2,24,34-37] These latter reviews should be consulted for more detailed discussions of aspects of plant
mitochondria not covered here. Given the level of interest
shown in the cyanide-resistant pathway, at least as judged
by the number of reviews in recent years, one is left won-
dering what to write about that has not already been dealt
with in a previous article. There is no satisfactory
answer to that question, but I will use the remainder of
this review to summarize what is currently known (and not

known) regarding the nature of the cyanide-resistant pathway
in general and the cyanide-resistant oxidase in particular.

THE CYANIDE-RESISTANT PATHWAY

Before proceeding, it would be useful to briefly consider the terminology used in referring to the cyanide-
resistant pathway and its corresponding oxidase. Because
the main cytochrome pathway is cyanide-sensitive, the
pathway which continues to operate in the presence of
added cyanide has been variously referred to as either the
cyanide-insensitive or the cyanide-resistant pathway. Both
are correct grammatically but the latter offers the slight
advantage of being more distinct from the term cyanide-
sensitive. This avoids the accidental mental juxtaposition
of the two similar terms (sensitive/insensitive) which can
occasionally occur. While the above point is admittedly
rather picky, the following one regarding the distinction
between the use of alternate versus alternative pathway,
is less so. Strictly speaking, the cyanide-resistant electron transfer chain represents an alternative pathway of
electron flow. There is no evidence that electrons are
partitioned between the cyanide-sensitive and resistant
pathways in true alternating fashion (i.e., one electron
down one and the next electron down the other) thus making
alternative (not alternate) pathway the title of choice.
Recent years have seen a gradual switch to the correct
terminology among workers in the field.

Occurrence of the Pathway

The observation that appreciable levels of a cyanide-
resistant component of respiration existed in certain plant
tissues can be traced back to the 1930s.[38] Much of the
early work focused on the extremely high respiratory rates
found in the spadix tissue of the "thermogenic aroids"
during flowering.[38] The link between this anomalous respiratory activity and normal aerobic metabolic processes
was provided in 1955 when James and Elliott[39] found that
mitochondria isolated from Arum maculatum spadices could
carry out a cyanide-resistant oxidation of TCA cycle substrates. Later Yocum and Hackett[40] demonstrated that Arum
mitochondria contained a normal, cyanide-sensitive cytochrome oxidase leading to the conclusion that an alternative,
cyanide-resistant oxidase must exist in these mitochondria.

After the initial insights provided in the mid 1950s,
only minimal progress was made toward a further understanding of the cyanide-resistant pathway until 1971. At that
time, Schonbaum et al.,[41] in a truly classic paper, reported
that aryl-substituted hydroxamic acids (Figure 3) specifically inhibited the cyanide-resistant pathway in isolated
plant mitochondria at concentrations which had no effect on
the cyanide-sensitive pathway. Prior to this report, no
really specific inhibitor of the alternative pathway was
known. After this report it was generally accepted that a
cyanide-resistant, hydroxamate-sensitive oxygen uptake
activity was diagnostic of the operation of the mitochondrial alternative pathway. While some caveats have appeared
in recent years concerning the use of the combination of
cyanide-resistance and hydroxamate-sensitivity to distinguish alternative pathway electron transfer (these will be
discussed later), this concept has, on the whole, been a
useful one and I doubt that the progress made to date toward
understanding this pathway would have been possible without
this class of inhibitors. While many different substituted
benzohydroxamic acids will specifically inhibit the alternative pathway, one, salicylhydroxamic acid (SHAM, Figure
3), has clearly become the hydroxamate of choice among
workers in the field, probably due to commercial availability as much as anything.

The alternative pathway has been shown to be present
in mitochondria isolated from a wide variety of higher plant
and algal sources. In 1975 Henry and Nyns[30] made a compila-
tion of species in which cyanide-resistant mitochondria had
been reported and numerous additional examples have appeared
since. It is also clear that the alternative pathway is not
found in all plant mitochondria and even the degree of
cyanide-resistance varies considerably among those plant
mitochondria which are cyanide-resistant. In some cases,

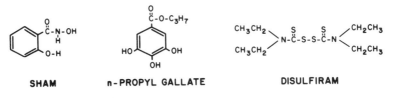

SHAM   n-PROPYL GALLATE   DISULFIRAM

Figure 3. Specific inhibitors of the alternative pathway
in plant mitochondria.

whether the mitochondria are cyanide-resistant or not depends on how the tissue used to obtain the mitochondria was treated prior to isolation. This has led to the general question of whether the alternative oxidase is a constitutive or inducible feature of plant mitochondria. An idea of the nuances associated with this question can be illustrated by considering the situation in potato tuber. Respiration in the intact tuber is not only resistant to but actually stimulated upon addition of cyanide.[42] However the oxygen uptake by mitochondria isolated from fresh potato tissue shows no appreciable cyanide-resistance.[43] If tuber slices are first allowed to "age" for 24 hours in aerated solution, mitochondria isolated from the resulting tissue show high levels of the cyanide-resistant, SHAM-sensitive pathway.[43] Laties and coworkers have examined this phenomenon and feel the loss of the alternative pathway in fresh tuber tissue is the result of a massive breakdown of membrane phospholipids which is induced upon slicing.[44] The restoration of cyanide-resistance with aging is associated with the synthesis of new membranes since inhibitors of either protein[45] or fatty acid[46] synthesis will prevent the appearance of the alternative pathway in aged tuber slices. Cyanide-resistant mitochondria can be isolated from fresh tuber tissue if the intact tubers are first held for 24-30 hours at room temperature in the presence of ethylene.[47-49] Day et al.[32] provide a more detailed overview of these results, but taken as a whole, they support the constitutive nature of the alternative pathway at least in potato tuber mitochondria.

Another illustration of the constitutive nature of the alternative pathway was provided in the recent studies of Rustin and coworkers.[29,50] They found that appreciable levels of cyanide-resistant (SHAM-sensitive) malate oxidation appeared in normally cyanide-sensitive potato tuber mitochondria upon addition of exogenous NAD. This activity was apparently due to stimulation of malate oxidation by malic enzyme in the mitochondrial matrix in the presence of added NAD. The resulting NADH was preferentially oxidized by the rotenone-resistant NADH dehydrogenase which in turn seemed to be closely linked to the alternative pathway. Results such as these point out how subtle the features governing the appearance of the alternative pathway in plant mitochondria can be.

In addition to higher plants and algae, the alternative pathway has been reported in a wide range of fungi including a large number of yeasts.[30]   Among animals, mitochondrial electron transfer is usually extremely sensitive to cyanide (0.1-1.0 mM) with rates being reduced to 1% or less of those found in its absence.[51]   However exceptions are known particularly among "lower" animals including Paramecium tetraurelia,[52] Acanthamoeba castellania[53] and several members of the brucei group of African trypanosomes.[54,55] The latter are flagellated protozoa causing African sleeping sickness (trypanosomiasis) in man and its counterpart, nagana, in domestic animals.   Trypanosomes present a particularly interesting story.[54]   The trypanosome spends part of its life cycle in an invertebrate vector (tsetse fly) and part in a suitable vertebrate host.   In the best characterized example, Trypanosoma brucei, the mitochondria are observed to degenerate considerably once the organism enters the bloodstream of the vertebrate host.   The trypanosome then carries out an aerobic form of glycolysis converting glucose to pyruvate at an extremely high rate.   The NADH generated during glycolysis is oxidized by a glycerol-3-phosphate dehydrogenase.   The resulting glycerol-3-phosphate moves into what remains of the mitochondria and is subsequently oxidized via a cyanide-resistant, SHAM-sensitive glycerol-3-phosphate oxidase.[55,56]   This oxidase will be discussed in more detail later.   So it appears that while cyanide-resistant respiration is most often associated with plants, the phenomenon is quite widespread among aerobic organisms.

Given that the alternative pathway is such a common feature of plant mitochondria, it is only natural to ask what role it might play in the life of the plant.   While that question has been placed outside the scope of this review, it should be noted that the physiological role of the alternative pathway in plants seems evident only in the case of thermogenesis that develops in some aroid spadices during flowering; a subject specifically reviewed by Meeuse.[38]   For a more thorough discussion of physiological aspects of the alternative pathway, particularly its possible role in plant respiratory metabolism, the reader should consult the recent reviews by Day et al.[32] and Lambers.[33]

Branch Point of the Pathway

Early studies of the alternative pathway led to speculation that either a flavoprotein or one of the mitochondrial b-type cytochromes was responsible for the observed
cyanide-resistant oxidase activity. However, in 1971,
Bendall and Bonner[57] showed that the mitochondrial b-type
cytochromes become fully reduced upon addition of the
inhibitor antimycin A while electron flow through the alternative pathway was unaffected. This localized the branch
point of the alternative pathway on the substrate side of
the cytochrome b pool. Because both succinate and NAD-
linked substrates such as malate are capable of supporting
cyanide-resistant electron flow, the results of Bendall and
Bonner further indicated that the alternative pathway
branches off the main pathway after electrons from complexes
I or II have entered a common pathway. This placed the
branch point in the vicinity of the ubiquinone pool. The
possibility that ubiquinone might serve as the branch point
was first approached experimentally by von Jagow and
Bohrer[58] using Neurospora crassa in which the alternative
pathway had been induced using chloramphenicol (see below).
They showed that the oxidation of exogenous NADH by both the
alternative and the cyanide-sensitive pathways was inhibited
in ubiquinone-depleted mitochondria and that both activities
were restored (40-50%) upon incorporation of ubiquinone back
into the depleted mitochondria.

More recently, Storey[59] carried out spectral studies of
the kinetics of oxidation of electron carriers in succinate-
reduced, carbon monoxide-saturated mitochondria from skunk
cabbage (Symplocarpus foetidus) spadices after pulsing with
oxygen. The use of carbon monoxide in these experiments to
block cytochrome oxidase overcame difficulties encountered
earlier using cyanide[60] or antimycin A.[61] Storey found that
ubiquinone and a specific flavoprotein, identified as FP_{ma}
(see ref. 2), were rapidly oxidized in the presence of CO
and that the rate of oxidation was decreased considerably
upon addition of m-chlorobenzohydroxamic acid (mCLAM).
Experiments using mitochondria isolated from potato tubers
which lacked the alternative pathway showed rates of
ubiquinone and flavopotein oxidation similar to those
observed with skunk cabbage mitochondria in the presence of
mCLAM plus CO. These results were interpreted in terms of
ubiquinone serving as the branch point of the alternative

pathway and FP_{ma} being the first component on the alternative pathway.

Electron paramagnetic resonance (EPR) spectroscopy has also been useful in helping to sort out the branch point of electrons onto the alternative pathway. A complex series of EPR signals centered around $g = 2.00$ appear in plant (and animal) mitochondria. These signals result from a spin-spin interaction between two stabilized ubisemiquinone species $(UQ^{-}UQ^{-})$.[62] Further, the rapid, low temperature relaxation of this split EPR signal suggests a close association between the semiquinone pair and the iron-sulfur center S-3 of complex II. In an initial study by Moore et al.,[63] it appears that some of the observations were the result of a marked effect of low concentrations of ethanol inhibiting the appearance of the split signal. This was corrected in a later study by Rich et al.[64] They found that appearance of the split signal was maximal under state 3 (+ADP) or uncoupled conditions while a marked decrease of the signal occurred in state 4 (+ATP) or upon addition of cyanide. The signal was inhibited entirely by addition of 1 mM SHAM. These results were taken to indicate that the species giving rise to the EPR signal (the semiquinone pair) was located on the main electron transfer pathway and in state 4 or the presence of cyanide, the semiquinone pair reached a more reduced steady state $((UQ^{-}UQ^{-} \rightarrow UQH_2UQH_2)$ leading to a loss of the EPR signal. Because these conditions (state 4 or +KCN) serve to shunt electrons onto the alternative pathway, the redox behavior of the semiquinone pair was felt to be consistent with that expected of the branch point of the alternative pathway as suggested earlier by Bahr and Bonner.[65,66]

The observation that addition of 1 mM SHAM to cyanide-treated mitochondria could abolish the split EPR signal supports the idea that the semiquinone pair serves as the branch point. However, care must be taken in interpreting results obtained utilizing this particular EPR signal. The concentration of SHAM required to inhibit the split EPR signal $(I_{50} = 200 \ \mu M)$ was much higher than that required to inhibit electron transfer through the alternative pathway $(I_{50} = 30 \ \mu M)$.[54] Further the split signal was abolished by 1 mM SHAM even in the absence of cyanide. The reasons for these discrepancies are not really clear. The presence of the semiquinone pair is likewise not required for the alternative pathway to operate. Concentrations of ethanol (0.1%) which have no effect on electron flow through the alternative pathway can completely abolish the split EPR signal in plant mitochondria.[64] In addition, the results of Rich et al.[64] do not rule out the possibility that the semiquinone pair is not the branch point per se, but is in rapid redox equilibrium with another pool (of ubiquinone?) which is the branch point. It should be remembered that the spin coupled ubisemiquinone pair is specifically associated with mediating electron flow from complex II while electrons from both succinate and internal NADH can be donated to the alternative pathway. In this regard, an EPR signal due to a stabilized ubisemiquinone species associated with complex III has been reported in mammalian mitochondria. It would be of interest in future work to determine the redox behavior of this semiquinone in plant mitochondria and the semiquinone's association, if any, with the alternative pathway.

Several lines of evidence support the concept that some component of the mitochondrial ubiquinone pool serves as the branch point for electrons off of the main onto the alternative pathway. However, the exact mechanism of branching is obscure because, contrary to the impression given in Figures 1 and 2, it is generally accepted that 1) electron flow through ubiquinone does not proceed in a strict linear fashion and 2) despite some kinetic evidence to the contrary,[67] ubiquinone probably does not act as a single homogeneous pool in mitochondria. Recent reviews of this subject by Gutman[3] and Trumpower[17] should be consulted for more detailed discussions. For the purposes of this review, it is sufficient to note that in 1975 Mitchell put forward the protonmotive Q cycle (Figure 4) as the mechanism of electron flow (and proton translocation) through ubiquinone in complex III.[68,69] The Q cycle itself seems to have been derived from earlier schemes.[70] Rich and Moore[71] have taken the view that ubiquinone serves as the branch point of the alternative pathway and integrated it with the Q cycle to speculate how electrons might branch off the main pathway. The basic concept put forward by Rich and Moore is shown in Figure 4. On the matrix side of the inner mitochondrial membrane, cytochrome b-560 reduces ubiquinone (UQ) to ubisemiquinone (UQ$^-$ or UQH$^{\cdot}$) in an antimycin A-sensitive step, and the dehydrogenases (complex I or II) drive the second half of the quinone redox reaction (UQH$^{\cdot}$→UQH$_2$). The

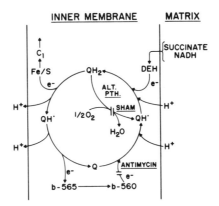

Figure 4. Protonmotive ubiquinone cycle and its proposed relation to the cyanide-resistant oxidative pathway (alt pth) in higher plant mitochondria. Adapted from Rich and Moore.[71]

alternative pathway must act to reverse the latter step or the pathway would be sensitive to antimycin A, which it isn't.

In spite of the seeming insight provided by schemes such as that described above and the accumulated evidence supporting ubiquinone as the site of the branch point onto the alternative pathway, there is still a considerable gap in our knowledge of the mechanism by which ubiquinone interacts with the alternative pathway. For example, external NADH is believed to feed electrons directly into the ubiquinone pool.[72] However, in many mitochondria the oxidation of external NADH is essentially cyanide-sensitive while that of succinate or internal NADH is cyanide-resistant.[31] Only in mitochondria from thermogenic tissue such as Arum spadices is external NADH oxidation consistently observed to be accessible to the alternative pathway.[73] This suggests there is more than one functional pool of quinone present in plant mitochondria and that the different pools are not 1) always able to achieve redox equilibrium or 2) equally accessible to the alternative pathway. In support of this Storey[74] has reported that the reduction of ubiquinone by external NADH showed biphasic kinetics while the quinone pool was reduced in a monotonic fashion when succinate was the electron donor. Huq and Palmer[73] found that partial extraction of ubiquinone from <u>Arum</u> mitochondria resulted in a preferential loss of the alternative pathway relative to the main pathway which they took to indicate the presence of two pools of quinone. As mentioned previously, Rustin et al.[29] found that internal NADH oxidized via the rotenone-resistant NADH dehydrogenase donated electrons preferentially to the alternative pathway while the rotenone-sensitive NADH dehydrogenase appeared to be linked more closely to the cyanide-sensitive pathway. This led them to propose that three separate quinone pools exist in plant mitochondria. One of the pools links the rotenone-resistant NADH dehydrogenase with the alternative pathway while the other two pools accept electrons from the rotenone-sensitive dehydrogenase and the external NADH dehydrogenase respectively and transfer them to the main pathway. The ability of these quinone pools to exchange electrons among one another varies in different plant mitochondria and this determines the pattern of cyanide-resistant oxygen uptake obtained using different electron donors.[29] It seems likely that separate quinone pools are present in both plant and animal mitochondria[3] and that a better understanding of how the quinone pools and the recently discovered protein-bound ubiquinone species[75,76] interact is needed before we can fully understand how electrons are shunted between the main and alternative pathways.

Inhibition of the Pathway

As noted earlier, probably the most significant advance in the past decade associated with the study of the alternative pathway was the finding that the pathway was specifically inhibited by substituted benzohydroxamic acids.[41] Prior to this report, Bendall and Bonner[57] had found that several metal binding agents (including thiocyanate, 8-hydroxyquinoline and α,α-dipyridyl) could inhibit the alternative pathway. The usefulness of these latter compounds was limited however in that they also inhibited the cyanide-sensitive pathway at concentrations only slightly higher than those required to inhibit the alternative pathway.[65] The hydroxamic acids, on the other hand, can completely inhibit the alternative pathway at concentrations (1-2 mM) where they have little or no effect on the main pathway. The concentration of hydroxamic acid required to inhibit the alternative pathway varies considerably with

both the hydroxamic acid used and the source of mitochondria.[41]

Schonbaum et al.[41] concluded that the hydroxamic acid
moiety (-(CO)NHOH, Figure 3) was required to observe inhibition. In their discussion of possible mechanisms by which
hydroxamates might act to inhibit the alternative pathway,
Schonbaum et al.[41] noted that these compounds are effective
bidentate chelators of the higher oxidation states of
several metals, the complex with iron (Fe^{3+}) being the best
known. But they did not favor the idea that hydroxamic
acids inhibited by binding to a transition metal at the
active site of the oxidase, feeling that cyanide should be
able to inhibit any active site accessible to a hydroxamate.
However they did observe that addition of m-iodobenzo-
hydroxamic acid led to enhancement of a g = 1.94-type EPR
signal associated with an iron-sulfur center in reduced
skunk cabbage submitochondrial particles. EPR studies of
the iron-sulfur centers in plant mitochondria since then
have revealed no signals unique to the alternative
pathway,[12-14] and none of the reported effects of hydroxamates on any mitochondrial EPR signals, in a manner which
would link the paramagnetic center to the alternative
oxidase, have stood the test of time.[41,63] Palmer[1] has also
pointed out that iron-sulfur proteins are generally resistant
to chelating agents except under denaturing conditions.
Nonetheless the notion that the hydroxamic acids inhibit the
alternative pathway via metal chelation, particularly to a
specific iron-sulfur center, has proved a compelling one[30,31]
and only in the absence of any supporting data has it gradu-
ally been rejected as a viable possibility.[24,77]

A second mechanism proposed by Schonbaum et al.[41] by
which hydroxamates could inhibit the alternative pathway
involved polyfunctional hydrogen bonding to the alternative
oxidase. This could stabilize one particular conformation
and prevent the oxidase from "turning over." This concept
came out of Schonbaum's work with horseradish peroxidase
which forms a stable complex with hydroxamic acids in a
competitive fashion with respect to the enzyme's phenolic
substrates.[78] Schonbaum et al regarded this latter theory
as the more likely of the two possibilities presented. It
should be pointed out that a basic assumption was made by
Schonbaum et al; this being that the alternative oxidase
itself is the site of inhibition by hydroxamates and not

some species a step or so removed from the oxidase. A
wealth of circumstantial evidence supports this assumption,
and I strongly suspect it is correct, but until the alternative oxidase has been characterized and its ability to
interact with hydroxamates confirmed, it would be worthwhile
to recognize that the hydroxamate binding site on the alternative pathway may not correspond to the alternative oxidase.

No additional work on the mechanism of inhibition of
the alternative pathway by hydroxamates appeared until 1978
when Rich et al.[79] surveyed a number of redox enzymes for
inhibition by hydroxamates. They found that in addition to
the alternative oxidase and horseradish peroxidase, the
copper-containing enzyme tyrosinase (polyphenol oxidase) was
also inhibited by hydroxamic acids. The latter two enzymes
differ from the alternative oxidase to the extent that they
are both inhibited by cyanide. These results clearly demonstrated that the alternative oxidase is not the only redox
enzyme sensitive to hydroxamates. More importantly, Rich et
al.[79] found that the inhibition of tyrosinase was competitive with respect to its phenolic reducing substrate and
essentially uncompetitive with respect to its oxidizing
substrate, molecular oxygen. These results prompted Rich et
al.[79] to suggest that a general mechanism of inhibition by
hydroxamates was as a competitive analogue of a phenolic
substrate for its binding site on the particular enzyme
being inhibited. How this concept relates to the alternative oxidase is outlined below.

By 1978, some very thorough studies had failed to turn
up any unique optical or EPR species which could be associated with the alternative pathway[13,14,64,80] and none of
the known electron carriers in the plant mitochondrial electron transfer chain behaved in a manner expected of the
alternative oxidase.[64,80] It was also generally accepted by
that time that electrons branched off the main chain at the
level of the ubiquinone pool as already discussed. Given
this background, two groups, Bonner and Rich[77] and Huq and
Palmer,[81] independently suggested that the alternative
oxidase was a quinol:oxygen oxidoreductase or, more simply,
a quinol oxidase. In Figure 4, the alternative pathway
would not represent a multicentered pathway but rather a
single cyanide-resistant, hydroxamate-sensitive oxidase.
Inhibition of the alternative oxidase by hydroxamates would
be analogous to the inhibition of tyrosinase; the hydroxamic

acid competing with reduced ubiquinone for its binding site on the alternative oxidase.[79] While the conceptual advance associated with the quinol oxidase theory has provided quite a boost to the field, a number of questions remain to be answered. For example, what species of reduced ubiquinone (UQ^-, UQH^-, UQH^-, UQH_2) serves as the substrate? The theory put forward by Rich and Moore[71] would suggest one of the fully reduced species (UQH^-, UQH_2) rather than ubisemiquinone but this remains to be proven. In line with an earlier consideration, can the bulk ubiquinone pool donate electrons to the alternative oxidase or is only a specific subpool of quinone accessible to it? Why don't hydroxamates interact with the quinol binding sites associated with electron flow through the main pathway (see Figure 4)? The answers to such questions need to be supplied before our understanding of the alternative pathway is complete.

If the alternative oxidase is a quinol oxidase which oxidizes reduced ubiquinone, it would formally be a p-diphenol:oxygen oxidoreductase. This is the same activity found in the enzyme laccase. Laccases are copper-containing enzymes which have been purified from both fungal and higher plant sources and are capable of oxidizing either ortho or para quinols and reducing oxygen to water in the process.[82,83] The distinction between laccase and polyphenol oxidase (o-diphenol:oxygen oxidoreductase) is often confusing in the literature, but the two are quite different enzymes as outlined in the review by Mayer and Harel.[82] While laccase is reported to be cyanide-sensitive,[82] the enzyme from Rhus latex requires long term incubations (>2 hrs) with very high concentrations of cyanide (20 mM) for inactivation.[84] Inactivation seems to be associated with the gradual removal of active site copper from the enzyme. So laccase is not cyanide-sensitive in the same sense that cytochrome oxidase or catalase is. It would be of interest to know if laccase is inhibited by hydroxamates. Based on the earlier findings with polyphenol oxidase (tyrosinase),[79] it is reasonable to expect that hydroxamates would inhibit laccase. If the alternative oxidase active site was identical to that found in laccase, EPR signals due to some of the copper species present in the enzyme should be seen under certain redox conditions[85] and no such signals have been reported. Paramagnetic Cu^{2+} of either Type 1 or Type 2 (both of which appear in laccase) is not a difficult species to see by EPR spectroscopy so the failure to observe such signals in plant mitochondria is significant.

The suggestion that the alternative oxidase might be a quinol oxidase led Rich and Bonner[86] and Huq and Palmer[81] to use reduced quinones as artificial electron donors to assay for the alternative oxidase. Several quinols, including duroquinol,[81] ubiquinol-1[86] and menaquinol,[86] were found to be oxidized in a cyanide-resistant, hydroxamate-sensitive manner by mitochondria from a number of different plant sources. Armed with an assay for the alternative oxidase, both groups set about to isolate the alternative oxidase and two preliminary communications on its isolation from Arum spadix mitochondria have appeared.[87, 88] In both cases, quinol oxidase activity was liberated from the mitochondrial membrane by detergent solubilization. Rich[88] used EPR spectroscopy to analyze a deoxycholate-solubilized, 100,000xg supernatant fraction which showed quinol oxidase activity. He found a copper signal around $g = 2$ and a small amount of $g = 4.3$ and $g = 1.93$ (iron-sulfur) iron. None of the EPR signals were observed to be redox active; presumably this meant that no changes were seen upon the addition of quinol under anaerobic conditions. A low level of flavin was reported to be present in this fraction. Huq and Palmer[87] also obtained a 100,000xg supernatant fraction by detergent (lubrol) solubilization, but they further purified their quinol oxidase activity using a series of DEAE-cellu-lose steps. Fluorescence measurements of this partially purified fraction indicated the presence of a flavin species, but the yellow color associated with the fraction was not bleached by addition of dithionite and was attri-buted to contaminating carotenoids. No significant EPR signals were observed and metal analysis showed copper to be the only metal present in appreciable amounts in the active fraction.

Unfortunately no further reports have appeared from either laboratory concerning quinol oxidase so the current status of these two preparations is a bit unclear. These initial results do support the idea that a specific, cyanide-resistant quinol oxidase is present in Arum mito-chondria and offer some hope that the nature of this enigmatic oxidase will be learned in the not too distant future. One caveat should be noted with regard to using the quinol oxidase assay. Extraneous metals, especially copper and to some extent iron, will catalyze the autoxidation of reduced quinones. These reactions are SHAM-sensitive but, unlike the oxidase activity referred to above, they are also inhibited by cyanide. In my laboratory we have found the problem of metal-catalyzed quinol oxidation is easily avoided by adding 1 mM EDTA to all reaction buffers.

While the studies cited above represent those most directly concerned with learning the nature of the alternative oxidase, other studies have appeared in recent years which have furthered our understanding of this pathway. In 1978, Parrish and Leopold[89] observed that lipoxygenase (linoleate:oxygen oxidoreductase) activity in imbibing soybean seed particles was inhibited by SHAM. Lipoxygenase is a nonheme iron-containing dioxygenase which catalyzes the addition of molecular oxygen across a double bond in polyunsaturated fatty acids having a cis,cis-1,4-pentadiene system.[90,91] Lipoxygenase has long been recognized as a cyanide-resistant enzyme so not only was this another example of a hydroxamate-sensitive oxidase, but also an example of one that, like the alternative oxidase, was cyanide-resistant. These findings and those of Rich et al.[79] discussed earlier served to negate the concept that any SHAM-sensitive, cyanide-resistant oxygen uptake activity in tissue respiration studies should automatically be attributed to the mitochondrial alternative pathway.

As a followup to the Parrish and Leopold[89] study, Siedow and Girvin[92] were led to consider what effect propyl gallate, a commercial antioxidant and a classic inhibitor of lipoxygenase,[91] (Figure 3), might have on the alternative pathway. They found that propyl gallate was a specific, reversible inhibitor of the alternative pathway in isolated mung bean mitochondria; showing half-maximal inhibition at concentrations (2-5 µM) an order of magnitude lower than those required of SHAM. Kinetic studies using both SHAM and propyl gallate indicated that the two compounds inhibit the alternative pathway in a mutually exclusive fashion. This suggested that the hydroxamates and propyl gallate inhibit the alternative pathway at the same or at least spacially similar sites. In a later study, Siedow and Bickett[93] utilized a series of propyl gallate analogues to identify the structural features required to inhibit the alternative pathway. They found the trihydroxy function on propyl gallate (Figure 3) was responsible for the observed inhibition.

They also showed that a single phenolic hydroxyl group was
the minimum structural feature required to specifically
inhibit the alternative pathway and that the phenolate
anion ($C_6H_5-O^-$) is probably the species that actually binds
to the mitochondria. These results rule out inhibition (by
propyl gallate) through bidentate metal chelation and are
compatible with the concept that propyl gallate (and SHAM)
inhibit the alternative pathway as competitive analogues of
some reduced form of ubiquinone (possibly UQH^- or $UQ^{\overline{\cdot}}$).

    Grover and Laties[94][95] recently reported another
specific inhibitor of the alternative pathway. Disulfiram
(tetraethylthiuram disulfide, Figure 3) irreversibly inhibited the alternative pathway in both red sweet potato
(Ipomoea batatas) and ethylene-treated white potato mitochondria with an I_{50} of 5-10 μM. The inhibition was not
mutually exclusive with hydroxamates indicating separate
sites of action for the two compounds. Because added thiols
could reverse (or prevent) the inhibition, Grover and Laties
suggested that disulfiram inhibition involved formation of a
mixed disulfide with one or more essential sulfhydryl groups
on the alternative pathway. Binding studies using [35]S-
labeled disulfiram indicated 4-6 nmoles of reactive sulfhydryl groups per mg protein were present in gradient purified potato tuber mitochondria whether the alternative
pathway was expressed (ethylene-treated) or not. The only
redox active component in plant mitochondria present at such
high levels is ubiquinone[96] and it is unlikely that disulfiram is reacting with ubiquinone per se as that might be
expected to inhibit the main pathway as well. It would
seem that as a greater number of specific inhibitors of the
alternative pathway are identified, they will provide more
latitude to workers attempting to probe the nature of the
alternative pathway and its associated oxidase.

Some Additional Considerations

    It was mentioned earlier that the alternative pathway
could be induced to appear in some plant mitochondria by
treating the parent tissue in various ways. It has also
been found that in some fungi, particularly yeasts, the conditions under which they are grown can lead to appearance
of the alternative pathway. Growing the yeast Saccharomy-
copsis lipolytica for 90 minutes in the presence of 1-2 mM
Fe^{3+} salts resulted in the appearance of a cyanide-resistant,

SHAM-sensitive oxygen uptake in isolated mitochondria that
was equivalent to the rate through the cyanide-sensitive
pathway.[97] Other conditions where the alternative pathway
can be induced to appear include inhibition of mitochondrial
protein synthesis (with chloramphenicol) in Neurospora
crassa,[98] limiting the level of copper in the growth medium
of Candida utilis,[99] or giving high concentrations of sulfate to Torulopsis utilis.[100] Unfortunately, none of these
observations has led to a better understanding of the nature
of the alternative pathway.

Two additional subjects which require brief mention are
the K_m of the alternative oxidase for oxygen and the product
of oxygen reduction. A number of tissue slice or whole cell
respiration studies have reported relatively high K_m values
for oxygen on the alternative pathway (10-25 μM or higher),
but these values are usually associated with diffusion-
limited steps as discussed by Solomos.[31] In isolated mitochondria, the K_m for oxygen of the alternative pathway is
found to be somewhat higher than that of cytochrome oxidase
but, nonetheless, quite low (<1-2 μM).[31,57] The determination of the product of oxygen reduction by the alternative
pathway has had a somewhat checkered history. For a long
time it was assumed that the alternative pathway brought
about the four electron reduction of oxygen to water[34] based
mainly on the apparent lack of formation of H_2O_2 by the
alternative pathway.[57] The finding that plant mitochondrial
preparations were contaminated with catalase led to a brief
period where H_2O_2 was elevated to the position of initial
product of the alternative pathway. Several lines of
experimentation have since led to the reestablishing of
water as the product of oxygen reduction.[88,101,102] Day et
al.[32] have covered the history of this subject quite well
so it will not be repeated here.

If water (and not peroxide or superoxide) is the product of oxygen reduction by the alternative oxidase, this
favors the presence of a metal at the oxidase active site
and goes against a flavin-containing oxidase or some autoxi-
dizable, protein-bound quinone species. There are no
examples in biology, to my knowledge, of a pure organic
species which mediates the reduction of molecular oxygen to
two molecules of water. Most organic reductants yield
either H_2O_2 or O_2^- upon reduction of oxygen[103] and to put 4
electrons on oxygen usually requires the presence of one or

more redox active metals.[104] Strictly speaking however the flavin-containing monooxygenases do bring about a 4 electron reduction of oxygen, but only one atom of the oxygen molecule ends up in H_2O; the second is incorporated into an organic species (R-OH).[105] Water is also the product of oxygen reduction by laccase, but the K_m for oxygen of laccase (and the other copper-containing oxidases) is much higher (>100 µM) than that found for the alternative oxidase.[82]

While attempts were being made to understand the alternative oxidase in higher plants, work was also being carried out on the cyanide-resistant, SHAM-sensitive glycerol-3-phosphate oxidase in the bloodstream form of trypanosomes, but the picture there is not much clearer. Fairlamb and Bowman isolated a particulate glycerophosphate oxidase from Trypanosoma brucei[106] and found it contained flavin (FAD), iron and copper. An EPR signal due to iron appeared at g = 4.3 but no EPR signal due to the copper could be generated in the final preparation even upon addition of ferricyanide. No mention was made of whether the quinone content of the oxidase preparation was measured or not. Fairlamb and Bowman[107] also showed that water, not H_2O_2, was the product of oxygen reduction. Some interest in the use of hydroxamates as trypanocidal drugs has been generated but with limited and confusing results in vivo (see ref. 108 for a summary). It would be useful to know if either propyl gallate or disulfiram is inhibitory to the trypanosome oxidase.

Trivial Explanations for the Pathway

Finally, it is worth discussing two recent series of studies which have ascribed the alternative pathway in isolated mitochondria to contamination either by lipoxygenase[109,110] or a nonmitochondrial NAD(P)H oxidase.[111,112] With respect to the former, Goldstein et al.[109] reported that wheat seedling mitochondria prepared by differential centrifugation contained a cyanide-resistant, hydroxamate-sensitive oxygen uptake but that the latter activity was lost if the mitochondria were further purified by centrifugation in Percoll density gradients. The latter procedure also served to remove contaminating lipoxygenase activity. Because both the alternative oxidase and lipoxygenase

activities were cyanide-resistant and SHAM (and propyl
gallate)-sensitive, Goldstein et al.[109] suggested in this
and a later report[110] that the alternative oxidase activity
in these mitochondria was due to the contaminating lipoxygenase. It was left unclear how lipoxygenase activity could
be linked to the oxidation of TCA cycle substrates such as
succinate or α-ketoglutarate, and no comparative titrations
of the two activities (lipoxygenase and alternative pathway)
with SHAM were reported. Half-maximal inhibition of lipoxygenase generally requires 4 to 5-fold higher concentrations
of SHAM than a comparable inhibition of the alternative
pathway (J. N. Siedow, unpublished observation).

Several lines of evidence can be marshalled which
indicate that alternative oxidase activity is not due to
lipoxygenase in many of the commonly utilized cyanide-
resistant mitochondria. A stoichiometric relation between
substrate oxidation and oxygen reduction through the alternative pathway has been demonstrated in a few cases.[29,101]
Purification of mitochondria by discontinuous sucrose
density gradient centrifugation leads to a marked reduction
in lipoxygenase levels[92] (similar to those found using
Percoll[109]), but with the exception of wheat, no loss or
drastic reduction in the level of the alternative pathway
has been reported.[113,114] Recent work in my laboratory has
also indicated that washed mitochondria isolated from the
spadices of two local aroids, Peltandra virginica (arrow
arum, Figure 5) and skunk cabbage (data not shown) have
appreciable levels of the alternative pathway but no detectable lipoxygenase activity (linoleic acid-stimulated oxygen
uptake). In addition, Miller and Obendorf[115] recently used
the inhibitor disulfiram to distinguish between the alternative pathway and lipoxygenase. They found that levels of
disulfiram sufficient to inhibit the alternative pathway
had no effect on contaminating lipoxygenase activity in isolated soybean mitochondria (Figure 6). Identical results
have subsequently been obtained in my laboratory using mung
bean mitochondria. It can be inferred in the two types of
potato tuber mitochondria used by Grover and Laties in their
original studies of disulfiram that the alternative pathway
there is not due to lipoxygenase.

While lipoxygenase is probably not associated with the
alternative pathway observed in those systems cited above,
Shingles and Hill[116] recently presented some interesting

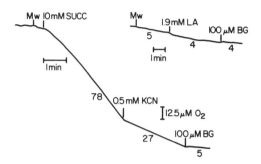

Figure 5. Alternative pathway and lipoxygenase activity in mitochondria isolated from spadices of Peltandra virginica (arrow arum). Tracing on the left shows the succinate-stimulated oxygen uptake in isolated mitochondria including the level of cyanide(KCN)-resistant, butyl gallate (BG)-sensitive pathway present. On the right, lipoxygenase activity is measured as the rate of linoleic acid(LA)-stimulated oxygen uptake and found to be absent in these mitochondria. Both activities were measured at pH 7.2 and are expressed below the traces as nmoles O_2 taken up/min/mg protein.

results which help clarify the situation in wheat coleoptile mitochondria and support the suggestion of Goldstein et al.[109] They found that substrate (malate or succinate)-linked, cyanide-resistant oxygen uptake did not correlate with substrate oxidation but rather with mitochondrial swelling. Substrate-induced swelling in some fashion pro-moted the release of free fatty acids from the mitochondrial membrane (possibly through stimulation of endogenous phos-pholipase activity) which then reacted with contaminating lipoxygenase leading to an apparent substrate-linked oxygen uptake. This idea was supported by showing that passive valinomycin-induced mitochondrial swelling stimulated a cyanide-resistant, propyl gallate-sensitive oxygen uptake in the absence of any added TCA cycle substrate (R. D. Hill, manuscript submitted). This cyanide-resistant activity in washed wheat mitochondria, while sensitive to SHAM and propyl gallate, was not inhibited by disulfiram. This

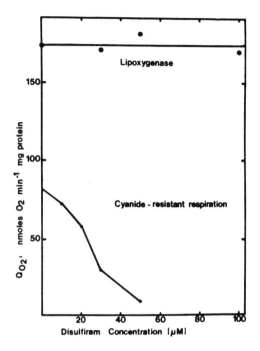

Figure 6. Effects of disulfiram on lipoxygenase activity
and on cyanide-resistant succinate oxidation in isolated
soybean mitochondria. Taken from Miller and Obendorf[115]
with permission.

differs from the mitochondria mentioned in the previous
paragraph, but is consistent with the alternative oxidase
activity being due to lipoxygenase in wheat coleoptile
mitochondria.

     The above results make it clear that some care must be
taken to establish that the linked oxidation of TCA cycle
substrates and reduction of oxygen is taking place when
studying the alternative pathway. This seems particularly
true when using mitochondria isolated from "nonstandard"
sources. Further, if such a "Rube Goldberg" series of
events can generate the cyanide-resistant oxidase activity
in wheat mitochondria, one cannot help but wonder if some

other combination of metabolic events might be operating to generate the alternative pathway that is found in mung bean or skunk cabbage mitochondria.

In a second series of studies, Ainsworth et al.[111,112] identified a SHAM-sensitive NAD(P)H oxidase activity in a nonmitochondrial membrane fraction as the source of the cyanide-resistant oxidase activity in baker's yeast (Saccharomyces cerevisae). While the intact yeast did show some cyanide-resistant oxygen uptake, baker's yeast mitochondria are not generally regarded as being cyanide-resistant[30] and no effort was made to show that they were in this study. The use of cyanide-resistant NADH oxidase activity to assay for the alternative oxidase led Ainsworth et al.[112] to a fraction which did show a limited SHAM-sensitivity (K_I = 370 μM). How a similar contaminating fraction in higher plant mitochondria could account for succinate-linked alternative oxidase activity isn't clear. Further, purified plant mitochondria which are essentially free of contaminating extraneous membrane fractions have been prepared using density gradient centrifugation[114] and, except for the wheat mitochondria discussed above, there has been no reported reduction in levels of the alternative pathway. There is no denying that a number of nonmitochondrial cyanide-resistant oxidases exist in plants (and fungi) which probably contribute to the overall rate of oxygen uptake in the intact cell. However there is no good reason, given the past 25 years of research, for believing that they are the source of the substrate-linked, alternative oxidase activity commonly found in isolated plant mitochondria.

It should be clear by now that in spite of the considerable effort that has been devoted to the study of the plant mitochondrial alternative pathway, we are still pretty much in the dark about most of the basic features of the pathway. Perhaps the conceptual advance provided by the quinol oxidase hypothesis will stimulate the experimental breakthrough so long overdue in this field. Only when the nature of the alternative oxidase becomes known will it be possible to understand either the role or the operation of this pathway in the overall metabolism in the cells of higher plants.

ACKNOWLEDGEMENTS

The preparation of this article and work in my laboratory was supported by a research grant from the NIH (GM 26095). I would like to express my thanks to Mark Bickett and all the students in my laboratory for their many stimulating thoughts on cyanide-resistant respiration and to S. D. Grover and R. D. Hill for making their manuscripts available prior to publication.

REFERENCES

1. Palmer, J. M. 1976. The Organization and Regulation of Electron Transport in Plant Mitochondria. Annu. Rev. Plant Physiol. 27: 133-157.
2. Storey, B. T. 1980. Electron Transport and Energy Coupling in Plant Mitochondria. In The Biochemistry of Plants, A Comprehensive Treatise (D. D. Davies, ed.). Vol. 2, Academic Press, New York, pp. 125-195.
3. Gutman, M. 1980. Electron Flux through Mitochondrial Ubiquinones. Biochim. Biophys. Acta 594: 53-84.
4. Moore, A. L., P. R. Rich and W. D. Bonner, Jr. 1978. Factors Influencing the Components of the Total Proton-motive Force in Mung Bean Mitochondria. J. Exp. Bot. 29: 1-12.
5. Fillingame, R. H. 1980. The Proton-Translocating Pumps of Oxidative Phosphorylation. Annu. Rev. Biochem. 49: 1079-1113.
6. Bonner, W. D., Jr. 1961. The Cytochromes of Plant Tissues. In Haematin Enzymes (J. E. Falk, R. Lemberg and R. K. Morton, eds). Pergamon Press, London. pp. 479-485.
7. Erecinska, M., D. F. Wilson, Y. Miyata. 1976. Mitochondrial Cytochrome b-c_1 Complex: Its Oxidation-Reduction Components and Their Stoichiometry. Arch. Biochem. Biophys. 177: 133-143.
8. Dutton, P. L., B. T. Storey. 1971. The Respiratory Chain of Plant Mitochondria, IX. Plant Physiol. 47: 282-288.
9. Lambowitz, A. L., W. D. Bonner, Jr. 1974. The b-Cytochromes of Plant Mitochondria. J. Biol. Chem. 249: 2428-2440.
10. Von Jagow, G., W. Sebald. 1980. b-Type Cytochromes. Annu. Rev. Biochem. 49: 281-314.

11. Lance, C., W. D. Bonner, Jr. 1968. The Respiratory
     Chain Components of Higher Plant Mitochondria.
     Plant Physiol. 43: 756-766.
12. Cammack, R., J. M. Palmer. 1973. EPR Studies of Iron-
     Sulfur Proteins of Plant Mitochondria. Ann. N.Y.
     Acad. Sci. 222: 816-823.
13. Cammack, R., J. M. Palmer. 1977. Iron-Sulfur Centers
     in Mitochondria from Arum maculatum Spadix. Biochem.
     J. 166: 347-355.
14. Rich, P. R., W. D. Bonner, Jr. 1978. The Nature and
     Location of Cyanide and Antimycin-resistant Respira-
     tion in Higher Plants. In Functions of Alternative
     Terminal Oxidases (H. Degan, D. Lloyd, G. C. Hill,
     eds.). Pergamon Press, Oxford, pp. 149-158.
15. Ohnishi, T., T. E. King. 1978. EPR and Other Proper-
     ties of Succinate Dehydrogenase. Methods in Enzymol.
     53: 483-495.
16. Rieske, J. S., R. E. Hanson, W. S. Zaugg. 1964.
     Studies on the Electron Transfer Chain, 58. J. Biol.
     Chem. 239: 3017-3022.
17. Trumpower, B. L. 1981. New Concepts on the Role of
     Ubiquinone on the Mitochondrial Respiratory Chain.
     J. Bioenerg. Biomembr. 13:1-24.
18. Prince, R. C., W. D. Bonner, Jr., P. A. Bershak. 1981.
     On the Occurrence of the Rieske Iron-Sulfur Cluster
     in Plant Mitochondria. Fed. Proc. Am. Soc. Exp.
     Biol. 40: 1667.
19. Roberts, H., W. M. Choo, S. C. Smith, S. Marzuki, A. W.
     Linnane, T. H. Porter and K. Folkers. 1978. The
     Site of Inhibition of Mitochondrial Electron Transfer
     by Coenzyme Q Analogues. Arch. Biochem. Biophys.
     191: 306-315.
20. Trumpower, B. L., J. G. Haggerty. 1980. Inhibition
     of Electron Transfer in the Cyt b-c_1 Segment of the
     Mitochondrial Respiratory Chain by a Synthetic
     Analogue of Ubiquinone. J. Bioenerg. Biomembr.
     12: 151-164.
21. Bowyer, J. R., G. V. Turney, A. R. Crofts. 1979. Cyt
     c_2 Reaction Center Coupling in Chromatophores of Rh.
     sphaeroides and Rh. capsulata. FEBS Lett.
     101: 207-212.
22. Bowyer, J. R., P. L. Dutton, R. Prince, A. R. Crofts.
     1980. Role of Rieske Iron-Sulfur Center as Electron
     Donor to Cytochrome c_2 in Rh. sphaeroides. Biochim.
     Biophys. Acta 592: 445-460.

23. Bowyer, J. R., B. L. Trumpower. 1980. Inhibition of the Oxidant-Induced Reduction of Cyt b by a Synthetic Analogue of Ubiquinone. FEBS Lett. 115: 171-174.

24. Palmer, J. M. 1979. The Uniqueness of Plant Mito-chondria. Biochem. Soc. Trans. 7: 246-252.

25. Coleman, J. O. D., J. M. Palmer. 1971. Role of Ca^{++} in the Oxidation of Exogenous NADH by Plant Mito-chondria. FEBS Lett. 17: 203-208.

26. Earnshaw, M. J. 1975. The Mechanism of K^{+}-Stimulated Exogenous NADH Oxidation in Plant Mitochondria. FEBS Lett. 59: 109-112.

27. Johnson, S. P., I. M. Moller, J. M. Palmer. 1979. The Stimulation of Exogenous NADH Oxidation in Jerusalem Artichoke Mitochondria by Screening of Charges on the Membranes. FEBS Lett. 108: 28-32.

28. Brunton, C. J., J. M. Palmer. 1973. Pathway for the Oxidation of Malate and Reduced Pyridine Nucleotide by Wheat Mitochondria. Eur. J. Biochem. 39: 283-291.

29. Rustin, P., F. Moreau, C. Lance. 1980. Malate Oxida-tion in Plant Mitochondria via Malic Enzyme and the Cyanide-insensitive Electron Transport Pathway. Plant Physiol. 66: 457-462.

30. Henry, M. F., E. J. Nyns. 1975. Cyanide-insensitive Respiration. An Alternative Mitochondrial Pathway. Sub-Cell. Biochem. 4: 1-65.

31. Solomos, T. 1977. Cyanide-resistant Respiration in Higher Plants. Annu. Rev. Plant Physiol. 28: 279-297.

32. Day, D. A., G. P. Arron, G. G. Laties. 1980. Nature and Control of Respiratory Pathways in Plants: The Interaction of Cyanide-resistant Respiration and the Cyanide-sensitive Pathway. In The Biochemistry of Plants, A Comprehensive Treatise (D. D. Davies, ed.). Vol. 2, Chapter 5, Academic Press, New York, pp. 197-241.

33. Lambers, H. 1980. The Physiological Significance of Cyanide-resistant Respiration in Higher Plants. Plant, Cell and Environ. 3: 293-302.

34. Ikuma, H. 1972. Electron Transport in Plant Respira-tion. Annu. Rev. Plant Physiol. 23: 419-436.

35. Bonner, W. D., Jr. 1973. Mitochondria and Plant Respiration. In Phytochemistry (L. P. Miller, ed.). Vol. 3, Van Nostrand-Reinhold, Princeton, pp. 221-261.

36. Moore, A. L., P. R. Rich. 1980. The Bioenergetics of Plant Mitochondria. Trends in Biochem. Sci. 5: 284-288.

37. Hanson, J. B., D. A. Day. 1980. Plant Mitochondria. In The Biochemistry of Plants, A Comprehensive Treatise (N. E. Tolbert, ed.). Vol. 1, Chapter 8, Academic Press, New York, pp. 315-358.

38. Meeuse, B. J. D. 1975. Thermogenic Respiration in Aroids. Annu. Rev. Plant Physiol. 26: 117-126.

39. James, W. O., D. C. Elliott. 1955. Cyanide-resistant Mitochondria from the Spadix of an Arum. Nature (London) 175: 89.

40. Yocum, C. S., D. P. Hackett. 1957. Participation of Cytochromes in the Respiration of the Aroid Spadix. Plant Physiol. 32: 186-191.

41. Schonbaum, G. R., W. D. Bonner, Jr., B. T. Storey, J. T. Bahr. 1971. Specific Inhibition of the Cyanide-insensitive Respiratory Pathway in Plant Mitochondria by Hydroxamic Acids. Plant Physiol. 47: 124-128.

42. Solomos, T., G. G. Laties. 1975. The Mechanism of Ethylene and Cyanide Action in Triggering the Rise in Respiration in Potato Tubers. Plant Physiol. 55: 73-78.

43. Dizengremel, P., C. Lance. 1976. Control of Changes in Mitochondrial Activities During Aging of Potato Slices. Plant Physiol. 58: 147-151.

44. Theologis, A., G. G. Laties. 1980. Membrane Lipid Breakdown in Relation to the Wound-induced and Cyanide-resistant Respiration in Tissue Slices. Plant Physiol. 66: 890-896.

45. Waring, A. J., G. G. Laties. 1977. Dependence of Wound-induced Respiration in Potato Slices on the Time-restricted, Actinomycin-sensitive Biosynthesis of Phospholipid. Plant Physiol. 60: 5-10.

46. Waring, A. J., G. G. Laties. 1977. Inhibition of the Development of Induced Respiration and Cyanide-insensitive Respiration in Potato Slices by Cerulenin and Dimethylaminoethanol. Plant Physiol. 60: 11-16.

47. Rychter, A., H. W. Janes, C. Frenkel. 1978. Cyanide-resistant Respiration in Freshly Cut Potato Slices. Plant Physiol. 61: 667-668.

48. Day, D. A., G. P. Arron, R. E. Christoffersen, G. G. Laties. 1978. Effect of Ethylene and Carbon Dioxide on Potato Metabolism. Plant Physiol. 62: 820-825.

metabolism in comparison to the massive metabolite fluxes found in the leaf peroxisomes or glyoxysomes. The following suggestions on the possible functions of these peroxisomes have been summarized.[12] First, the organelles may catabolize metabolites like glycolate and urate that cannot be oxidized by the mitochondria. Second, during peroxisomal oxidation, no useful energy in the form of ATP or reduced pyridine nucleotides is produced, and thus the peroxisomes may provide a mechanism for the disposal of excess cellular energy. Finally, since some of the H_2O_2-producing oxidases have a low affinity for oxygen, the peroxisomes may protect the cells from oxygen toxicity under high oxygen tension.

Although no massive metabolite flux appears to occur in the unspecialized peroxisomes, the organelles may still perform unique and important functions that have yet to be discovered. Many cellular processes do not involve massive metabolite fluxes but nevertheless serve indispensible physiological roles, such as hormone production. Unless and until such physiological roles are discovered, this type of peroxisomes will remain to be called unspecialized peroxisomes.

Glyoxysomes

Glyoxysomes are present in the fatty tissues of seeds during germination (Figure 1). They were the first plant peroxisomes to be isolated and studied biochemically.[4] The organelles were discovered as novel particles containing enzymes of the glyoxylate cycle, and were thus termed glyoxysomes. Subsequent studies indicate that the glyoxysomes are also the exclusive sites of the β-oxidation of storage fatty acid, and of the characteristic peroxisomal enzymes including catalase, glycolate oxidase, and urate oxidase.[16,17] The organelles from the castor bean endosperm have been studied most intensively, and studies on the glyoxysomes from other oil seeds reveal many similarities and only minor differences.[14,18]

During the germination of oil seeds, triacylglycerols in the storage tissue are rapidly mobilized. Sucrose is a major product which is transported to the growing embryonic axis. The long gluconeogenic pathway involves many enzymes in several distinct subcellular compartments, including the lipid bodies, the glyoxysomes, the mitochondria, and the

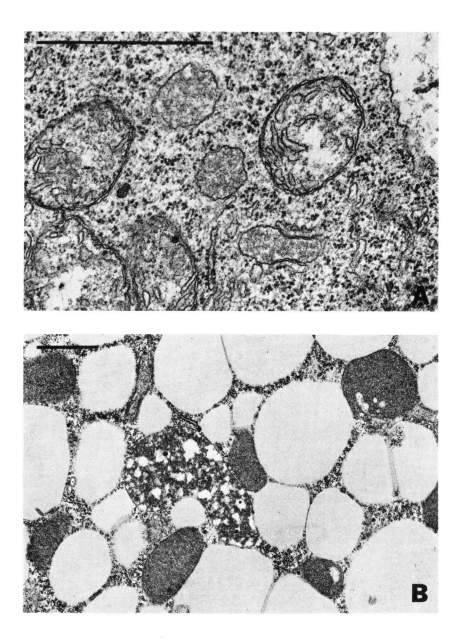

Figure 1A and 1B (Legend, see page 94)

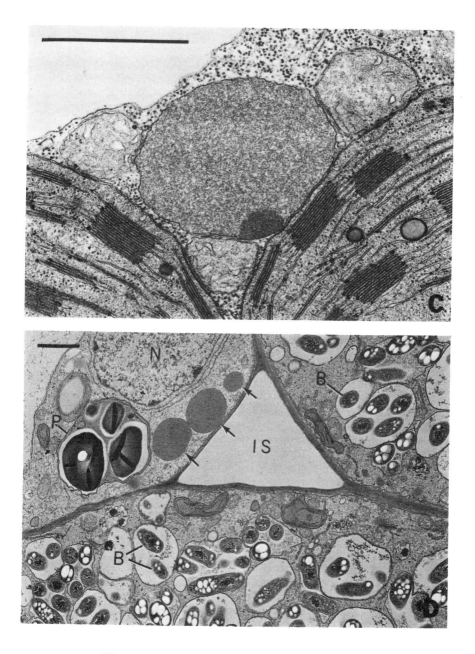

Figure 1C and 1D (Legend, see page 94)

Figure 1.   Electron micrographs of the four types of peroxisomes in higher plants listed in Table 1.   A. Portion of a
bean root tip cell showing three unspecialized peroxisomes
(courtesy of R. N. Trelease).   B. Portion of a cotyledon
cell of 4-day-old dark-grown cucumber seedling, showing
several glyoxysomes among numerous lipid bodies.   Small
pockets of cytoplasm have invaginated into some of the
glyoxysomes (courtesy of R. N. Trelease).   C. Portion of a
tobacco leaf cell showing a leaf peroxisome and several
mitochondria and chloroplasts (by S. E. Frederick, courtesy
of E. H. Newcomb).   D. Portion of infected and uninfected
cells of a soybean nodule, showing several microbodies
(peroxisomes involved in ureide metabolism) in the uninfected cell.   At bottom and upper right are infected cells
with numerous bacteroids (B) in vacuoles.   Abbreviations:
IS, inter-cellular space; N, nucleus; P, plastid with
several starch grains (Newcomb and Tandon, 1981;[41] courtesy
of E. H. Newcomb).   The bar on each micrograph represents
1 µm.

---

cytosol[19] (Figure 2).   The gluconeogenic pathway begins with
lipid hydrolysis.   Uncertainties exist as to which subcellular compartment contains the lipase for triacylglycerol
hydrolysis.[20,21]   Strong evidence indicates that the
mechanism may be different in various oil seeds.[22]   Irres-
pective of whether the lipase in the glyoxysomes or some
other organelles is responsible for triacylglycerol
hydrolysis, the fatty acid released is metabolized exclusively in the glyoxysomes.   It is first activated by a
synthetase to fatty acyl CoA.[23]   Fatty acyl CoA enters the
β-oxidation sequence consisting of four enzymes.   The first
enzyme, fatty acyl CoA oxidase, catalyzes an oxidative
reaction using molecular oxygen and producing H_2O_2.   The
glyoxysomes contain an active catalase which can destroy
the toxic H_2O_2 as soon as it is formed.   The product of
β-oxidation is acetyl CoA which is then channeled through
the glyoxylate cycle.   The glyoxylate cycle then converts
two molecules of acetate into one molecule of succinate
which is the end product of glyoxysomal metabolism.   The
succinate produced is reduced to malate by the TCA cycle in
the mitochondria.   Malate is decarboxylated to phosphoenolpyruvate which is then converted to hexose by the enzymes
of reverse glycolysis in the cytosol.

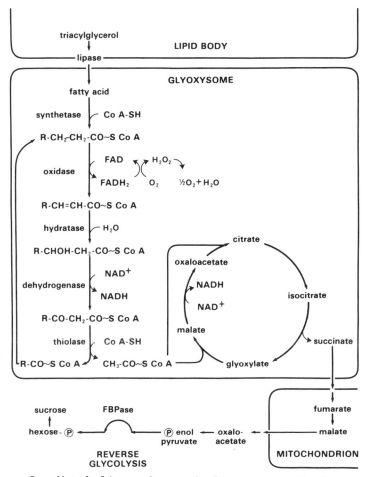

Figure 2.   Metabolic pathway of gluconeogenesis from storage triacylyceroles in the castor bean endosperm.   An elaborate electron shuttle system to account for the reoxidations of NADH generated in the glyoxysomes has been proposed recently.[24]   Malate instead of succinate would leave the glyoxysomes to be converted to sucrose directly in the cytosol.   Succinate would go to the mitochondria as indicated in this Figure, and the malate formed by the TCA cycle would return to the glyoxysomes to replenish the deprived malate. By this mechanism, the malate of the glyoxylate cycle would carry carbon skeleton as well as the reducing equivalent of NADH out of the glyoxysomes.

Two reactions in the glyoxysomes reduce NAD to NADH.
For each sequence of the β-oxidation, one NADH is generated,
and for each turn of the glyoxylate cycle, another NADH is
produced (Figure 2). The glyoxysomes do not have the
capacity to re-oxidize NADH, and they contain only a catalytic amount of NAD/NADH.[24] Apparently, the NADH is re-
oxidized outside the glyoxysomes, presumably in the mitochondria. An electron shuttle system across the glyoxysomal
membrane has been proposed to account for the reoxidation
of NADH. The proposal involves the participation of two
glyoxysomal enzymes, malate dehydrogenase and aspartate-α-
ketoglutarate transaminase, either directly (Figure 3) or
indirectly through an elaborate shuttling of metabolites
between the glyoxysomes and the mitochondria.[25,26] The
proposal requires further studies, especially using
isolated glyoxysomes in an intact form during the assay
period.

The glyoxysome is the best-studied plant peroxisome.
Many important enzymes of the β-oxidation and the glyoxylate
cycle are associated with the organelle membrane, but the
significance of such an association is unknown.[27] [28] The
biogenesis of glyoxysomes at the molecular level is being
actively pursued, and this topic is excluded in this article
dealing with subcellular metabolism. The ontogenic inter-
relationship of glyoxysomes and leaf peroxisomes in the
fatty cotyledons of seeds during germination has been critically reviewed recently,[29] and will not be discussed in this
article.

Leaf Peroxisomes

Leaf peroxisomes are present in green leaves (Figure
1), green photosynthetic cotyledons of seedlings, and
perhaps also in other photosynthetic tissues. The major
physiological process that the leaf peroxisomes participate
in is photorespiration.[5,12] Glycolate generated in the
chloroplasts by RuBP carboxylase and phosphoglycolate phosphatase (or by other mechanisms) is oxidized to glyoxylate
exclusively in the leaf peroxisomes. In addition, the leaf
peroxisomes contribute heavily to the subsequent metabolism
of glyoxylate.

The leaf peroxisomes contain a highly active glycolate
oxidase which catalyzes the oxidation of glycolate in the

presence of O_2 to produce glyoxylate and H_2O_2. The organelles also possess an active catalase which presumably can
destroy the toxic H_2O_2 as soon as it is formed. The glyoxylate generated may (1) undergo transamination in the proposed glycolate pathway (Figure 3), (2) be oxidized by H_2O_2
non-enzymatically to form CO_2 and formate,[30] (3) be oxidized
to oxalate catalyzed by glycolate oxidase,[31] (4) return to
the chloroplasts for oxidative decarboxylation,[32] or (5)
return to the chloroplasts for reduction to glycolate.[12]

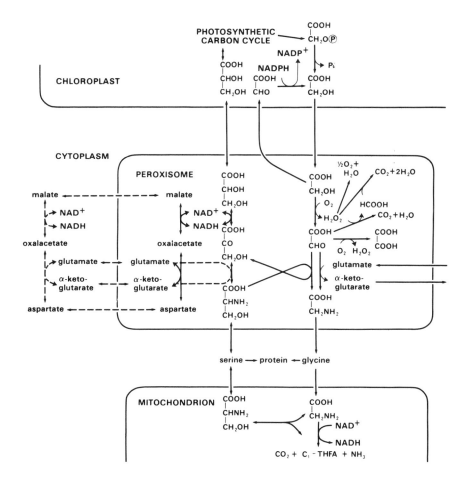

Figure 3. Glycolate metabolism in photorespiration (adopted
from Tolbert, 1981[39]).

It seems likely that a large portion of the glyoxylate
goes through the proposed glycolate pathway when the
availability of amino group for glyoxylate transamination
is not limited (Figure 3).

Besides glycolate oxidase and glyoxylate transaminase,
the leaf peroxisomes also contain active enzymes of serine-
hydroxypyruvate transaminase and hydroxypyruvate reductase
as participants of the entire glycolate pathway[12] (Figure 3).
Whereas the conversion of glycolate to glycine is not
readily reversible, the interconversion between serine and
glycerate (sometimes termed the glycerate pathway) is freely
reversible. During photorespiration, the metabolite flow in
the peroxisomes is from glycolate to glycine and from serine
to glycerate. The conversion of glycerate to serine is
operational in situations other than photorespiration (e.g.,
in the dark) for the generation of serine, glycine, and
methylene tetrahydrofolic acid which are essential metabolites for cellular metabolism.

During photorespiration, the complete glycolate pathway
generates NAD and the reverse glycerate pathway produces
NADH. Since an NADH oxidase has not been found in the
peroxisomes, an electron shuttle system has been proposed[12]
(Figure 3). The presence of a highly active peroxisomal
NAD-malate dehydrogenase, which has no other known metabolic
function in the organelles, and a peroxisomal aspartate
α-ketoglutarate transaminase suggests that they may be
involved in the shuttle mechanism. Malate and aspartate,
but not oxaloacetate, would be the components of the shuttle.
This proposed shuttle system is similar to the proven system
in the mitochondria[26] and the proposed system in glyoxysomes
(see previous section).

Peroxisomes are also present in the leaves of CAM
species and in both the leaf mesophyll and bundle sheath
tissues of C_4 species, as revealed by electron microscopy.[33]
The question has been raised as to whether these peroxisomes
are indeed the leaf-type peroxisomes as those characterized
in C_3 species, or just the unspecialized peroxisomes. The
peroxisomes from a few CAM and C_4 species have been isolated
and shown to contain catalase and/or glycolate oxidase.[14]
However, these two enzymes are also present in other types
of peroxisomes. It is likely that the peroxisomes in the

leaves of CAM species are similar to the leaf peroxisomes in C_3 species, since photorespiration does occur in CAM species under favorable environments.[34] C_4 species do not carry out overt photorespiration.[35] Some C_4 species even have no "internal" photorespiration, and they possess dimorphic chloroplasts. Since the mesophyll cells do not contain RuBP carboxylase and thus do not generate glycolate, there is no purpose of having leaf-type peroxisomes in the tissue. Similarly, the functional role of leaf-type peroxisomes in the bundle sheath of those C_4 species having a highly reduced glycolate metabolism (i.e., internal photorespiration) is questionable. Indirect evidence supports the suggestion that the peroxisomes in the leaf mesophyll cells of Sorghum bicolor[36] and some other C_4 species[35] are not the leaf-type peroxisomes. Further studies on the peroxisomes in the leaves of C_4 species are warranted.

Peroxisomes in Ureide Metabolism

The ureides, allantoin and allantoic acid, are important metabolites for nitrogen transport in the phloem of some plant species. In species of the genera Persea, Acer, Platanus, and Aesculus, the xylem saps contain allantoin and allantoic acid as the major forms of nitrogen, representing 10-99% of the total nitrogen.[37] For the transport of organic nitrogen from the nodules to the shoots of symbiotic nitrogen-fixing legumes, some species like Vicia and Pisum translocate asparagine whereas other species like Glycine and Vigna transport ureides as the major nitrogen carriers.[38]

The metabolic pathway of ureides is shown in Figure 4. In some ureotelic animals, several of the enzymes, including xanthine oxidase (dehydrogenase), urate oxidase, and allantoinase, are localized in the peroxisomes.[39] In plants, low activities of urate oxidase are also present in the various types of peroxisomes, including the glyoxysomes,[40] the leaf peroxisomes, and the unspecialized peroxisomes[15] (Table 2). In some of these peroxisomes, allantoinase is also present, although the activity is again very low.[40] The peroxisomes and their possible content of ureide metabolic enzymes have only been studied recently in the tissues of those plant species that utilize ureides as nitrogen transporting metabolites.

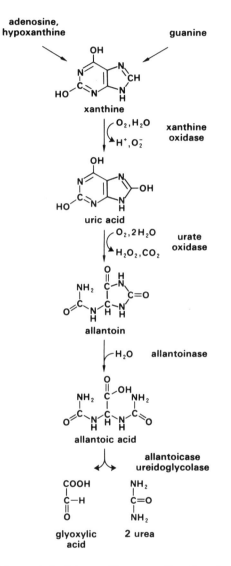

Figure 4. Ureide metabolic pathway. In the last reaction, allantoic acid is first metabolized to ureidoglycolic acid by allantoinase, and then to glyoxylate and urea by ureidoglycolase. Other catabolic reactions of allantoic acid also exist in microorganisms.[37]

In the nodule extracts of Glycine (soybean) and Vigna (cowpea), substantial activities of xanthine dehydrogenase, urate oxidase, and allantoinase are present.[38] In soybean nodules, large and proliferous peroxisomes are present in the uninfected root cells, adjacent to the infected root cells which contain small or degenerated peroxisomes[41] (Figure 1). In the nodules, catalase and urate oxidase are localized in the peroxisomes whereas xanthine oxidase and allantoinase are restricted to the cytosol and the microsomes, respectively.[42] Presumably, allantoin is produced by urate oxidase in the peroxisomes in the uninfected nodule cells. These peroxisomes contain a high activity of urate oxidase which is at least several fold higher than that in any other types of plant peroxisomes (Table 2). In the leaves, little allantoinase is present; catalase, together with only a trace amount of urate oxidase, exists in the peroxisomes, whereas allantoinase occurs in the microsomes.[43] Apparently, unlike the peroxisomes in some ureotelic animals, soybean peroxisomes contain only a small part of the ureide metabolic pathway. Nevertheless, the metabolism of ureide in relation to nitrogen transport should be recognized as an important physiological role carried out by plant peroxisomes (Table 1). More studies on the peroxisomes in tissues where ureides are metabolized likely will be forthcoming.

PEROXISOMES IN FUNGI AND ALGAE

The peroxisomes in gymnosperms, lower tracheophytes, and bryophytes will not be discussed in detail in this article, since substantial biochemical information is lacking. A comprehensive review of the literature will be available (Huang, Moore, and Trelease, Plant Peroxisomes, Academic Press, in preparation). Briefly, microbodies have been observed with the electron microscope in green and non-green cells of diverse species. In a few cases, cytochemistry has revealed their content of catalase. Glyoxysomes from the megagametophyte of pine seeds have been studied biochemically, and they appear to be similar to those in castor bean. Glyoxysomes are likely to occur also in fern spores during germination, although unequivocal evidence is lacking. In green tissues, leaf-type peroxisomes possessing glycolate oxidase are likely to exist, but cell fractionation experiments have not been performed.

Taken together, the peroxisomes in lower tracheophytes and bryophytes appear to be similar to those in flowering tracheophytes.

In fungi and algae, peroxisomes have been observed by electron microscopy in diverse species, regardless of whether they are autotrophs or heterotrophs growing on diverse media. The literature is prolific, and contains considerable confusion. A review has not appeared which puts into proper perspectives the current knowledge of the physiological aspect of these peroxisomes.

There are technical difficulties that lead to confusion in the study of peroxisomes in microorganisms. The algal or fungal cells are difficult to break gently in a way that permits the fragile organelles to be isolated intact. This difficulty may be a major cause for the contradictory reports of enzyme localization. Some confusion also has arisen because the investigators may not be aware of the dependence of the physiological role of the peroxisomes on various factors, especially the carbon source for growth. Catalase and isocitrate lyase have been the two enzymes commonly employed as markers for the peroxisomes and the glyoxylate cycle, respectively. Interpretation of experimental results may be erroneous when investigators were not aware of certain properties of these two enzymes. For example, in Saccharomyces cerevisiae, two forms of catalase exist, one localized in the peroxisomes and the other in the vacuole.[44] Using catalase indiscriminately as a marker for peroxisomes in yeast may therefore lead to confusion. Isocitrate lyase is an inducible enzyme when microorganisms are grown on acetate, ethanol, or alkane, and those microorganisms grown on glucose or sucrose contain a very low activity of the enzyme. It is of little value to isolate peroxisomes from glucose-grown microorganisms and use isocitrate lyase as a means to assess the existence of glyoxysome-like organelles. Furthermore, in Neurospora, two isozymic forms of isocitrate lyase are present; one is produced constitutively in a low amount and the other is an inducible enzyme.[45,46]

In fungi and algae, at least six distinct types of peroxisomes can be distinguished according to their metabolic roles (Table 1). It is anticipated that peroxisomes with other metabolic roles will be identified in the

future, such as peroxisomes in those microorganisms grown
on glycolate or urate as the sole carbon or nitrogen source.

Peroxisomes of Unknown Function-"Unspecialized" Peroxisomes

Peroxisomes are present in most if not all fungi and
algae when examined with the electron microscope.[47,48]
Their content of catalase in most cases has been confirmed
by cytochemical methods or by direct enzyme assay on organelles isolated by sucrose gradient centrifugation.[14,47]
Whereas the peroxisomes in microorganisms grown under
certain environments perform unique physiological functions
(Table 1), the peroxisomes in other microorganisms, such as
those grown on glucose, sucrose, or lactose as the sole
carbon source, do not carry out any known, active metabolic
role. In these microorganisms, the number of peroxisomes
per cell, or the number compared with other organelles, is
generally small. No physiological function has been proposed for these peroxisomes other than performing in the
general detoxication of H_2O_2. In essence, they are similar
to the "unspecialized" peroxisomes of higher plants. Again,
caution should be made that the term "unspecialized" refers
to the lack of known special physiological role, and the
status may be altered with new information.

Glyoxysome-like Peroxisomes in Heterotrophic Fungi and Algae

When heterotrophic fungi or algae are grown on certain
carbon sources that are related to the metabolites along the
gluconeogenic pathway from triacylglycerols in oil seeds
(Figure 5), glyoxysome-like peroxisomes are induced. These
carbon sources, including alkane, fatty alcohols, fatty
acids, ethanol, and acetate, can enter the known metabolic
pathway directly or through initial modification. Also, the
spores of some fungi contain storage lipid which is
mobilized during or following spore germination, and the
situation is similar to the germination of oil seeds. The
overall metabolic pathway in heterotrophic fungi and algae
is quite similar to the early portion of gluconeogenesis in
oil seeds, but with some differences.

A major difference between the glyoxysomes in oil seeds
and the glyoxysome-like peroxisomes in fungi is that the
fungal peroxisomes contain only the two key enzymes of the
glyoxylate cycle, malate synthase and isocitrate lyase, and

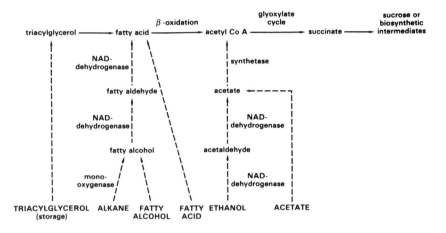

Figure 5. Metabolism of lipid derivatives by fungi and heterotrophic algae. The metabolic reactions carried out by the microorganisms produce initial products that can be channeled through the known gluconeogenic pathway in oil seeds (indicated in the upper portion).

thus have to carry out the glyoxylate cycle by sharing enzymes with the TCA cycle in the mitochondria. Among the heterotrophic algae, only Euglena gracilis is known to have enzymes of the complete glyoxylate cycle in the peroxisomes. Because only a few studies have succeeded in the isolation of peroxisomes, it is not known whether Euglena is unique among the heterotrophic algae.

Peroxisomes in fungal spores with storage triacyl-glycerol. The spores of some fungi contain storage lipid in the lipid bodies. Following germination, the number of peroxisomes increases, and many of them are situated adjacent to the lipid bodies.[49] The peroxisomes have been isolated from three fungal species, Blastocladiella emersonii,[50] Entophylyctis variabilis,[51] and Botryodiplodia theobromae.[52] The peroxisomes isolated from these fungal spores contain isocitrate lyase and malate synthase of the glyoxylate cycle, as well as catalase. The other glyoxylate cycle enzymes, citrate synthase and malate dehydrogenase, are present in the mitochondria and not in the peroxisomes.

In B. theobromal, enoyl-CoA hydratase is present only in
the mitochondria, and about 80% of thiolase and β-hydroxy
acyl CoA dehydrogenase activities are present in the mitochondria and 20% in the peroxisomes. It thus seems that
the peroxisomes in the spores of the above three fungal
species are not as sophisticated as those in oil seeds, and
have to cooperate with the mitochondria in carrying out the
conversion of fatty acid to acetate and then to succinate.

Peroxisomes in fungi and algae grown on alkane, fatty
alcohol and fatty acid. Some fungi, algae, and bacteria can
use alkane, fatty alcohol, or fatty acid as the sole carbon
source for growth. There is an increasing interest in this
research area because of its applicability to converting
petroleum by-products into consumable food stuffs, cleaning
accidental oil spills, and resolving corrosion in jet
aircraft fuel systems due to fungal growth.

The initial enzymatic steps for the utilization of
alkane in fungi (Candida tropicalis and Cladosporium resinae)
and bacteria involve the following reactions.[53-55]

Monooxygenase
$R-CH_3$ → $R-CH_2OH$ → $R-CHO$ → $R-COOH$ → β-oxidation
alkane   alcohol   aldehyde   acid

The first enzyme is a monooxygenase requiring both O_2 and
NAD as electron acceptors. The second and third enzymes
presumably utilize NAD as the electron acceptor. The
resultant fatty acid is oxidized by the β-oxidation system
to form acetate which will be used for energy production
or assimilation.

The involvement of peroxisomes in the metabolism of
hydrocarbon derivatives has been well-studied in Candida
tropicalis.[56-58] The peroxisomes and related enzymes are
induced when the yeast is grown on alkane, and the peroxisomes become more numerous than the mitochondria. The monooxygenase for the oxidation of alkane to alcohol is located
in the microsomes. About one-third of the cellular fatty
alcohol dehydrogenase, fatty aldehyde dehydrogenase, and
fatty acyl CoA synthetase, are present in the mitochondria
and microsomes, presumably for membrane lipid biosynthesis.
The remaining two-thirds of the above three enzyme activities are present in the peroxisomes, which also contain the

characteristic peroxisomal enzymes of catalase, urate
oxidase, and D-amino acid oxidase. In addition, the peroxisomes contain all of the β-oxidation enzymes as well as
isocitrate lyase and malate synthase of the glyoxylate
cycle. The other three glyoxylate cycle enzymes, (citrate
synthase, malate dehydrogenase, and aconitase) are absent
from the peroxisomes. The mitochondria do not contain the
β-oxidation enzymes, but possess enzymes of the complete
TCA cycle. Thus, the peroxisomes and the mitochondria must
cooperate in operation of the glyoxylate cycle, with metabolites shuttling between the two organelles (Figure 6).
In addition, both organelles contain carnitine

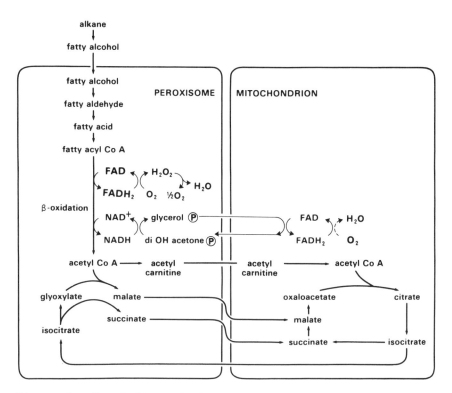

Figure 6. Metabolism of alkane by peroxisomes and mitochondria in Candida tropicalis. Succinate, malate, or
other TCA cycle intermediates may be drained away for other
cellular metabolism.

acetyltransferase, which may provide a mechanism for trans-
fering acetate units generated by the β-oxidation system in
the peroxisomes to the mitochondria for the citrate synthase
reaction (Figure 6). Most of the cellular NAD-glycerol
phosphate dehydrogenase and FAD-glycerol phosphate dehydrogenase are restricted to the peroxisomes and the mitochondria, respectively. The two enzymes can provide an
electron shuttle system for transferring the NADH generated
by the β-oxidation in the peroxisomes to the mitochondria
for oxidation, using glycerol phosphate and dihydroxyacetone
phosphate as the vehicles. Such an electron shuttle system
is known to occur between the cytosol and the mitochondria
in mammalian cells.[39]

Although having many similarities, the Candida peroxisomes differ from the oilseed glyoxysomes in several
aspects. The yeast peroxisomes do not contain a complete
glyoxylate cycle, but possess carnitine acetyltransferase
and NAD-glycerol phosphate dehydrogenase. These two enzymes
are unknown in the peroxisomes of higher plants but exist
in mammalian peroxisomes.[39] The presence of fatty alcohol
dehydrogenase and fatty aldehyde dehydrogenase in the yeast
peroxisomes is also unique.

In heterotrophic Euglena grown on hexanoate, the isolated peroxisomes contain fatty acyl CoA synthetase, two of
the β-oxidation enzymes examined, and the complete glyoxylate cycle enzymes.[59] The Euglena peroxisomes are thus very
similar to the glyoxysomes of higher plants.

Peroxisomes in fungi and algae grown on ethanol or
acetate. When fungi and heterotrophic algae use ethanol or
acetate as the sole carbon source for growth, an increase
in the activities of enzymes that can utilize ethanol or
acetate occurs (Figure 5). Ethanol is first converted to
acetate by ethanol dehydrogenase and acetaldehyde dehydrogenase. Acetate is activated to acetyl CoA which is then
channeled through the glyoxylate cycle. The two dehydrogenases and acetyl CoA synthetase (Figure 5) are induced by
ethanol and acetate.[59,60]

The peroxisomes isolated from several fungal species,
including Neurospora species,[61] Saccharomyces cerevisiae,[62]
Coprinus lagopus,[63] and Aspergillus tamarii,[64] generally
contain the characteristic peroxisomal enzymes, including

catalase, urate oxidase, and D-amino acid oxidase. In
addition, the peroxisomes contain malate synthase and isocitrate lyase but not the other three enzymes (citrate
synthase, aconitase, and malate dehydrogenase) of the
glyoxylate cycle. Thus, the peroxisomes and the mitochondria must work together for operation of the glyoxylate
cycle. The mechanisms of cooperation, such as the shuttles
and source of acetyl CoA in either organelle, are unknown.
Whether or not the mechanisms are similar to those of
Candida grown on alkane (Figure 6) remains to be seen.

        In heterotrophic algae grown on acetate or ethanol,
the subcellular localization of the glyoxylate cycle may be
species-dependent. In Euglena gracilis,[65] the isolated
peroxisomes contain acetyl CoA synthetase and the complete
glyoxylate cycle enzymes (all five enzymes except aconitase
were studied). No catalase is present in the peroxisomes or
in the algal extract, as is the case in autotrophic Euglena
(see next section). On the contrary, peroxisomes isolated
from Polytomella caeca contain catalase and urate oxidase,
but malate synthase and isocitrate lyase appear only in the
soluble fraction.[66,67] The uniqueness of Euglena or
Polytomella among diverse algal species in peroxisomal
metabolism is unknown.

Peroxisomes and Photorespiration in Autotrophic Algae

        In many ways, algae carry out photorespiration in a
similar way as higher plants.[68] However, differences exist.
Under normal environmental conditions, algae carry out a
lower rate of photorespiration. This is advantageous since
the loss of CO_2 from new photosynthate is reduced. The
reduction has been postulated to be due to a more favorable
internal low O_2/CO_2 ratio in the aqueous environment.
Besides the reduction in overt photorespiration, differences
in the biochemical pathway also exist.

        A major difference in the biochemistry of glycolate
metabolism between algae and higher plants lies in the
glycolate oxidizing enzyme.[69,70] Some algae are similar to
higher plants (tracheophytes and bryophytes) in possessing
glycolate oxidase that transfers electrons directly to O_2 to
form H_2O_2. Other algae contain a related enzyme, glycolate
dehydrogenase, that does not transfer electrons directly to

oxygen but transfers electrons to an artificial electron acceptor like 2,6-dichlorophenol indophenol. The native electron acceptor is unknown, and NAD is not active.

The glycolate oxidase reaction generates H_2O_2, and thus requires the presence of catalase to decompose the toxic H_2O_2. On the other hand, the glycolate dehydrogenase reaction does not produce H_2O_2, and there is no apparent need for catalase. Those algae possessing glycolate dehydrogenase have greatly reduced catalase activity, generally 10-50% of the levels on a protein or chlorophyll basis found in higher plants and algae possessing glycolate oxidase.[70]

In algae containing glycolate dehydrogenase, the capacity of the forward glycolate pathway is limited by the low activity of glycolate dehydrogenase which on a per unit of photosynthesis basis is generally less than 10% of glycolate oxidase activity in C_3 plants.[68,70] Excess glycolate which is produced and cannot be metabolized by the limited activity of glycolate dehydrogenase will be excreted to the environment. It is estimated that up to 10% of the photosynthate in the form of glycolate is excreted to the environment under normal CO_2 and O_2 concentrations in the atmosphere. Since the forward glycolate pathway may not generate enough serine, glycine, and methylene tetrahydrofolic acid for other cellular metabolism, algae generally utilize the reversible portion of the glycolate pathway in converting glycerate to serine and glycine more heavily than do higher plants (Figure 3).

Among algae possessing glycolate oxidase, peroxisomes have been isolated from two species. In Spirogyra sp., the peroxisomes contain catalase, glycolate oxidase, and hydroxypruvate reductase.[71] In Chlorella vulgaris, the peroxisomes also contain catalase and glycolate oxidase.[72] However, there is a conflicting report claiming that glycolate dehydrogenase[70] instead of glycolate oxidase is present in this algal species. In Klebsormidium flaccidum, a cytochemical method also establishes that glycolate oxidase is localized in the peroxisomes.[73] From the limited existing information, it appears that the peroxisomes in algae that contain glycolate oxidase are similar to the leaf peroxisomes in higher plants.

Among those algae that contain glycolate dehydrogenase, the peroxisomes from autotrophic Euglena gracilis have been studied most intensively.[74-76] This algal species does not contain detectable catalase activity. The peroxisomes contain all the particulate hydroxypyruvate reductase and serine-glyoxylate transaminase, and only a portion of the particulate glutamate glyoxylate transaminase, aspartate-α-ketoglutarate transaminase, and malate dehydrogenase. Glycolate dehydrogenase is present in about equal proportions in the peroxisomes and the mitochondria. Thus, the Euglena peroxisomes, like the leaf peroxisomes, can metabolize glycolate to glycine. Since the peroxisomal glycolate dehydrogenase reaction does not generate H_2O_2, the absence of catalase in the peroxisomes is not deleterious. The mitochondria, with sufficient levels of glycolate dehydrogenase, glutamate-glyoxylate transaminase, and the glycine oxidation system, have the capacity to metabolize glycolate to glycine, serine, and CO_2. The mitochondrial glycolate dehydrogenase is linked to the electron transport system and eventually to O_2 as the final electron acceptor. The reversible interconversion between glycerate and serine catalyzed by hydroxypyruvate reductase and serine glyoxylate transaminase is restricted to the peroxisomes (Figure 3). Should Euglena oxidize glycolate through the mitochondrial glycolate dehydrogenase, useful metabolic energy is generated through the electron transport system. In this way, the mitochondrial oxidation system is advantageous over the peroxisomal glycolate oxidase of higher plants and other algae. The reason that Euglena possess glycolate dehydrogenase in two subcellular compartments is unknown. The peroxisomal enzyme, but not the mitochondrial enzyme, is induced under conditions that are favorable for photorespiration, and it was thus suggested to be the one in operation during photorespiration.[76,77]

The data on the subcellular localization of glycolate dehydrogenase in Euglena are convincing and were reproduced by several laboratories.[74,76,78] Limited information is available on other algae possessing glycolate dehydrogenase. In Cylindrotheca fusiformis, Nitzschia alba[79] and Chlamydomonas reinhardtii[80], the enzyme is present in the mitochondria; whether or not it is also present in the peroxisomes is unclear.

In summary, the subcellular location of glycolate dehydrogenase and other enzymes of the glycolate pathway in glycolate dehydrogenase-containing algae is poorly understood. Whereas in _Euglena_ the enzyme is undoubtedly present in the mitochondria, the role of the mitochondrial enzyme in photorespiration has not been resolved. Whether or not algae other than _Euglena_ also contain glycolate dehydrogenase in the peroxisomes is unclear. The enzymes for the interconversion between glycerate and serine are restricted to the peroxisomes in _Euglena_, but their subcellular location in other algae is unknown. With such a great taxonomical diversity in algae, one should be cautious in extending the information obtained from _Euglena_ to other algae without further investigations. The evolution of the mitochondrial glycolate dehydrogenase, the peroxisomal glycolate dehydrogenase, and the peroxisomal glycolate oxidase in algae deserves further attention.

Peroxisomes in Methanol Metabolism

Methanol can be used by bacteria and some fungi as the sole carbon source for growth.[81] As shown by radioactive tracer experiments and enzyme studies, fungal cells metabolize methanol initially by the following reaction:

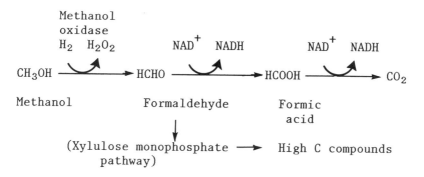

Formaldehyde is metabolized either to CO_2 for energy production or to more complex carbon compounds by the xylulose monophosphate pathway.

Methanol oxidase from _Candida boidinii_ possesses the characteristics of a peroxisomal H_2O_2-producing flavoenzyme.[82] The enzyme has not been reported to be present in the peroxisomes of higher plants. In yeast, it is

localized, together with catalase, in the peroxisomes in
Candida boidinii,[83] Hansenula polymorpha,[84] and Kloeckera
sp.[85] Since the methanol oxidase reaction generates H_2O_2,
the localization of the oxidase together with catalase in
the peroxisomes is logical. Formaldehyde dehydrogenase and
formate dehydrogenase are present in the cytosol.[83] The
peroxisomes can be seen as conspicuous and numerous
organelles in methanol-grown yeast.[86] Many of the peroxisomes contain crystalline cores with clearly visible
crystalline lattice.

Some bacteria utilize methanol by a serine pathway
involving isocitrate lyase.[81] This pathway does not appear
to be operative in yeast. In methanol-grown yeast cells,
isocitrate lyase activity is very low, and its activity does
not correlate well with those of methanol-induced enzymes or
with peroxisome formation.[87] It seems that methanol metabolism in yeast does not involve the glyoxylate cycle or a
part of the cycle.

Peroxisomes in Oxalate Synthesis

Many plant pathological fungi secrete oxalic acid as
well as cell wall hydrolytic enzymes during infection. The
process has been studied extensively in Sclerotium rolfsii,
a soilborn fungus that is pathologic to a wide range of
plant species around the world. The oxalic acid creates a
lower pH environment for the acidic pectinases, and at the
same time loosens the cell wall by chelating the cell wall
calcium.[88] Oxalic acid production also occurs when the
fungus is grown on extracted host cell wall polysaccharides
or other artificial media, but is greatly reduced when
grown in simple carbon source at a low pH or unbuffered
media.

Oxalic acid is produced by the following enzymatic
reactions:

$$\text{Isocitrate} \xrightarrow[\text{lyase}]{\text{Isocitrate}} \text{Glyoxylate} \xrightarrow[\text{dehydrogenase}]{\overset{\displaystyle NAD^+ \quad NADH}{\text{Glyoxylate}}} \text{Oxalate}$$

The properties of NAD-glyoxylate dehydrogenase were studied
in a cell-free extract of S. rolfsii.[89] The enzyme utilizes

NAD^+ as the electron acceptor. The Km value for glyoxylate is 1.2×10^{-4} M, and the enzyme does not act on glycolate, ethanol, lactate, and pyruvate at 1.3 mM or higher concentrations. The specificity of the enzyme toward the substrate and electron acceptor suggests that it is not a lactate dehydrogenase, a xanthine oxidase, or a glycolate oxidase that can also oxidize glyoxylate to oxalate. This enzyme has not been described in other organisms.

During the active infection period, numerous peroxisomes are present in the fungal hypha.[48] The two enzymes, isocitrate lyase and glyoxylate dehydrogenase, together with catalase, are localized in the peroxisomes.[90] No malate synthase was detected in the peroxisomes, suggesting that isocitrate lyase does not participate in the glyoxylate bypass. Whether isocitrate is derived from the TCA cycle in the mitochondria, or the peroxisomes also contain citrate synthase and aconitase for isocitrate production, is unknown.

The two key enzymes, isocitrate lyase and glyoxylate dehydrogenase, are not involved in H_2O_2 production. Their localization in the peroxisomes, however, is in accord with the idea that peroxisomes compartmentalize reactions involving glyoxylate, preventing this reactive metabolite from being diverted into other undesirable reactions. Although spinach and a few other plant species produce and accumulate substantial amounts of oxalate, glyoxylate dehydrogenase is absent in isolated spinach leaf peroxisomes.[91]

Peroxisomes in Amine Metabolism

Yeasts of many genera can use various amines as the sole nitrogen source for growth.[92-94] These amines include aliphatic mono-, di-, and triamines of short carbon chains, and simple aromatic amines. Monoamines are metabolized by the following reaction:

$$RCH_2NH_2 + H_2O \quad \xrightarrow[\text{amine oxidase}]{O_2 \quad H_2O_2} \quad RCHO + NH_3$$

Ammonia is assimilated to organic nitrogen compounds, whereas RCHO is oxidized further in the presence of

formaldehyde dehydrogenase. In general, yeasts cannot
utilize amines as the sole nitrogen and carbon souces,
indicating that they do not assimilate RCHO. The above
amine oxidation may be involved in the catabolism of ethanolamine and choline from phospholipids. Yeasts can grow on
ethanolamine or choline as the sole nitrogen source.
Ethanolamine is an active substrate of amine oxidase.
Choline is converted to trimethylamine and subsequently to
monomethylamine, which is the oxidized to formaldehyde in
the presence of amine oxidase.

     Amine oxidase from Aspergillus niger is a copper
enzyme that is most active toward aliphatic monoamines of
C_3 to C_6.[93] Amine oxidase, together with catalase, is
localization in the peroxisomes in Candida utilis and
Hansenula polymorpha.[94] The two enzymes and formaldehyde
dehydrogenase are induced after the yeasts have been trans-
fered to media containing amines as the sole nitroge source.
Under this growth condition, the number of peroxisomes is
enhanced also.

PERSPECTIVES AND FUTURE RESEARCH ON PEROXISOMAL METABOLISM

     The lack of information on the permeability of peroxi-
somal membrane is a serious deficiency in our understanding
of peroxisomal metabolism. In the study of chloroplasts
and mitochondria, many properties, activities, and control
mechanisms of the organelles are altered drastically after
the intact organelles have been deliberately broken. The
peroxisomes have been studied biochemically for more than
10 years, and yet intact organelles have never been subjected to in vitro experimentation. This deficiency is
mainly because the peroxisomes have been separated from
other organelles only by equilibrium sucrose gradient centrifugation. The isolated peroxisomes are obtained in 54%
(w/w) sucrose (density 1.25 g/cm3). All attempts to dilute
the sucrose solution to a workable density and viscosity by
various methods have resulted in complete organelle breakage,
presumably due to osmotic lysis. If the peroxisomes are
studied in 54% sucrose, the high viscosity and density of
the solution prevents a meaningful interpretation of experimental results. Investigators have been proposing electron
shuttle systems for the oxidation/reduction of NADH/NAD
generated in the peroxisomes. Yet, we still do not know if
NADH/NAD can pass through the peroxisomal membrane. The

membrane may only serve the purpose of packaging the enzymes in a compartment, or it may also regulate metabolism by controlling the flow of metabolite in and out of the organelles.

Besides the permeability of the peroxisomal membrane, several other major research needs in the metabolism in peroxisomes in higher plants have been discussed throughout the text. The outstanding problems include identification of the initial substrate for the glyoxysomes and the significance of the association of the major enzymes with the glyoxysomal membrane. The metabolic fate of glyoxylate generated from glycolate in the leaf peroxisomes should be determined. The study on ureide metabolism in peroxisomes in relation to nitrogen transport has just begun and deserves more efforts.

The knowledge on peroxisomal metabolism in fungi and algae has been fragmentary, and the lack of a complete picture is understandable in view of the great diversity of genetic background among fungi or algae. Sufficient information is available to make only a few generalizations. The study of peroxisomes in yeast grown on alkanes has been most fruitful, and their function has been clearly established. The technique of isolation of peroxisomes from autotrophic _Euglena_ has been successful, and the information obtained so far is encouraging. The _Euglena_ appears to be a good system for investigating the functions of peroxisomes and mitochondria in autotrophic algae with glycolate dehydrogenase.

A knowledge of the functions of microbial peroxisomes may provide new insights into several important applied problems. The information on glyoxysome-like peroxisomes which metabolize alkanes and other fatty components, as well as methanol-oxidizing peroxisomes, may be applicable to the petroleum industry alone, or linked also to food production. The algal peroxisomes and photorespiration should be further studied with the goal of deriving biological systems for harvesting the energy of the sun. The work on oxalate synthesis may be applicable to the control of phytopathological fungi. The application of known specific inhibitors of isocitrate lyase or glyoxylate dehydrogenase should control fungal growth without harming the host, since both enzymes are absent in the major tissues of the host plants.

ACKNOWLEDGMENT

    A large portion of this article is condensed from a
chapter in "Plant Physiology Monograph: Plant Peroxisomes"
under the authorship of A. Huang, T. Moore, R. Trelease, to
be published by Academic Press. I wish to thank R. Trelease
and T. Moore for their excellent assistance in the literature search as well as their very critical but constructive
review on the original manuscript of the Monograph chapter.
I also thank Drs. E. H. Newcomb and R. Trelease for supply-
ing excellent electron micrographs and Dr. R. Moreau for
reviewing the manuscript critically. My research on peroxisomes in the past was supported by the National Science
Foundation, and currently is being supported by the USDA
Competitive Research Grant Program (5901-0410-9-0317-0).

REFERENCES

1. Thorton, R. M., K. V. Thimann. 1964. On a crystal-
    containing body in cells of the oat coleoptile.
    J. Cell Biol. 20: 345-350.
2. Mollenhauer, H. H., J. D. Morre, A. G. Kelly. 1966.
    The widespread occurrence of plant cytosomes resemb-
    ling animal microbodies. Protoplasma 62: 44-52.
3. Frederick, S. E., E. H. Newcomb, E. L. Vigil, W. P.
    Wergin. 1968. Fine-structural characterization of
    plant microbodies. Planta 8: 229-252.
4. Breidenbach, R. W., H. Beevers. 1967. Association of
    glyoxylate cycle enzymes in a novel subcellular
    particle from castor bean endosperm. Biochem.
    Biophys. Res. Comm. 27: 462-469.
5. Tolbert, N. E., A. Oeser, T. Kisaki, R. H. Hageman,
    R. K. Yamazaki. 1968. Peroxisomes from spinach
    leaves containing enzymes related to glycolate
    metabolism. J. Biol. Chem. 243: 5179-5184.
6. de Duve, C., P. Baudhuin. 1966. Peroxisomes (micro-
    bodies and related particles). Physiol. Rev.
  • 46: 323-357.
7. Beevers, H., R. W. Breidenbach. 1974. Glyoxysomes.
    Methods Enzymol. 31A: 565-571.
8. Tolbert, N. E. 1974. Isolation of subcellular organ-
    elles of metabolism on isopycnic sucrose gradients.
    Methods Enzymol. 31A: 734-746.

9. Vigil, E. L., G. Wanner, R. R. Theimer. 1979. Isolation of plant microbodies. In Plant Organelles (E. Reid, ed.). Ellis Horwood limited, Chichester. pp. 89-102.

10. Huang, A. H. C. 1981. Isolation and subfractionation of glyoxysomes. Methods Enzymol. 72: 783-790.

11. de Duve, C. 1969. Evolution of the peroxisomes. Ann. N.Y. Acad. Sci. 168: 369-381.

12. Tolbert, N. E. 1980. Microbodies - peroxisomes and glyoxysomes. In The Biochemistry of Plants (P. K. Stumpf, E. E. Conn, eds.). Academic Press, New York. Vol. 1, Chapter 9, pp. 359-388.

13. Cooper, T. G., R. P. Lawther. 1973. Induction of the allantoin degradative enzymes in Saccharomyces cerevisiae by the last intermediate of the pathway. Proc. Natl. Acad. Sci. USA. 70: 2340-2344.

14. Gerhardt, B. 1978. Microbodies/peroxisomen pflanzlicher Zellen. Cell Biology Monographs Volume 5. Wien: Springer. 283 pp.

15. Huang, A. H. C., H. beevers. 1971. Isolation of microbodies from plant tissues. Plant Physiol. 48: 637-641.

16. Cooper, T. C., H. Beevers. 1969. β-oxidation in glyoxysomes from castor bean endosperm. J. Biol. Chem. 244: 3514-3520.

17. Hutton, D., P. K. Stumpf. 1969. Characterization of the oxidation systems from maturing and germinating castor bean seeds. Plant Physiol. 44: 508-516.

18. Huang, A. H. C. 1975. Comparative studies on glyoxysomes from different fatty seedlings. Plant Physiol. 55: 870-874.

19. Beevers, H. 1969. Glyoxysomes of castor bean endosperm and their relation to gluconeogenesis. Ann. N. Y. Acad. Sci. 168: 313-324.

20. Ory, R. L. 1969. Acid lipase of the castor bean. Lipids. 4: 177-185.

21. Muto, S., H. Beevers. 1974. Lipase activities in castor bean endosperm during germination. Plant Physiol. 54: 23-28.

22. Huang, A. H. C., R. A. Moreau. 1978. Lipases in the storage tissues of peanut and other oil seeds during germination. Planta 141: 111-116.

23. Cooper, T. G. 1971. The activation of fatty acids in castor bean endosperm. J. Biol. Chem. 246: 3451-3455.

24. Mettler, J. J., H. Beevers. 1980. Oxidation of NADH in glyoxysomes by a malate-aspartate shuttle. Plant Physiol. 66: 555-560.

25. Cooper, T. C., H. Beevers. 1969. Mitochondria and glyoxysomes from castor bean endosperm. J. Biol. Chem. 246: 3451-3455.

26. Hanson, J. B., D. Day. 1980. Plant mitochondria. In The Biochemistry of Plants (P. K. Stumpf, E. E. Conn, eds.). Vol. 1, Academic Press, New York. pp. 315-358.

27. Huang, A. H. C., H. Beevers. 1973. Localization of enzymes within microbodies. J. Cell Biol. 58: 379-389.

28. Bieglmayer, C., J. Graf, H. Ruis. 1973. Membranes of glyoxysomes from castor bean endosperm. Enzymes bound to purified membrane preparations. Eur. J. Biochem. 37: 553-562.

29. Beevers, H. 1979. Microbodies in higher plants. Annu. Rev. Plant Physiol. 30: 159-193.

30. Halliway, B. 1974. Oxidation of formate by peroxisomes and mitochondria from spinach leaves. Biochem. J. 138: 77-85.

31. Richardson, K. E., N. E. Tolbert. 1961. Oxidation of glyoxylic acid to oxalic acid by glycolic acid oxidase. J. Biol. Chem. 231: 1280-1284.

32. Zelitch, I. 1972. The photooxidation of glyoxylate by envelop-free spinach chloroplasts and its relation to photorespiration. Arch. Biochem. Biophys. 150: 698-707.

33. Frederick, S. E., E. H. Newcomb. 1971. Structure and distribution of microbodies in leaves of grasses with and without CO_2 photorespiration. Planta 96: 152-176.

34. Osmond, C. B., J. A. M. Holtum. 1980. Crassulacean acid metabolism. In The Biochemistry of Plants (P. K. Stumpf, E. E. Conn, eds.). Vol. 8, Academic Press, New York. pp. 283-328.

35. Edwards, G. E., S. C. Huber. 1981. The C_4 pathway. In The Biochemistry of Plants (P. K. Stumpf, E. E. Conn, eds.). Vol. 8, Academic Press, New York. pp. 238-281.

36. Huang, A. H. C., H. Beevers. 1972. Microbody enzymes and carboxylases in sequential extracts from C_4 and C_3 leaves. Plant Physiol. 50: 242-248.

37. Thomas, R. J., L. E. Schrader. 1981. Ureide metabolism in higher plants. Phytochemistry 20: 361-371.

38. Rawsthorne, S., F. R. Minchin, R. J. Summerfield, C. Cookson, J. Coombs. 1980. Carbon and nitrogen metabolism in legume root nodules. Phytochemistry 19: 341-355.

39. Tolbert, N. E. 1981. Metabolic pathways in peroxisomes and glyoxysomes. Annu. Rev. Biochem. 50: 133-157.

40. Theimer, R. R., H. Beevers. 1971. Uricase and allantoinase in glyoxysomes. Plant Physiol. 47: 246-251.

41. Newcomb, E. H., S. R. Tandon. 1981. Uninfected cells of soybean root nodules: ultrastructure suggests key role in ureide production. Science 212: 1394-1396.

42. Hanks, J. F., N. E. Tolbert, K. R. Schubert. 1981. Localization of enzymes of ureide biosynthesis in peroxisomes and microsomes of nodules. Plant Physiol. 68: 65-69.

43. Hanks, J. F., K. R. Schubert, N. E. Tolbert. 1981. Subcellular localization of allantoinase in soybean leaves. Plant Physiol. 67: 28S.

44. Susani, M., P. Zimniak, F. Fessl, H. Ruis. 1976. Localization of catalase A in vacuoles of Saccharomyces cerevisiae: evidence for the vacuolar nature of isolated "yeast peroxisomes." Hoppe-Seyler's Z. Physiol. Chem. 357: 961-970.

45. Armitt, S., C. F. Roberts, H. L. Kornberg. 1970. The role of isocitrate lyase in Aspergillus nidulans. FEBS Letters 7: 231-234.

46. Sjogren, R. E., A. H. Romano. 1967. Evidence for multiple forms of isocitrate lyase in Neurospora crassa. J. Bacteriol. 93: 1638-1643.

47. Silverberg, B. A. 1975. An ultrastructural and cytochemical characterization of microbodies in the green algae. Protoplasma 83: 269-295.

48. Maxwell, D. P., V. N. Armentrout, L. B. Graves, Jr. 1977. Microbodies in plant pathogenic fungi. Annu. Rev. Phytopathol. 5: 119-134.

49. Powell, M. J. 1978. Phylogenetic implications of the microbody-lipid globule complex in zoosporic fungi. Biosystems 10: 167-180.

50. Mills, G. L., E. C. Cantino. 1975. The single microbody in the zoospore of Blastocladiella emersonii is a "Symphyomicrobody". Cell Differentiation 4: 35-44.

51. Powell, M. J. 1976. Ultrastructure and isolation of glyoxysomes (microbodies) in zoospores of the fungus Entophylyctis sp. Protoplasma 89: 1-27.

52. Armentrout, V. N., D. P. Maxwell. 1981. A glyoxysomal role for microbodies in germinating conidia of Botryodiplodia theobromae. Exper. Mycol. 5: In Press.

53. LeBeault, J. M., B. Roche, Z. Duvnjak, E. Azoulay. 1970. Alcool et aldehyde-deshydrogenases particulaires de Candida tropicalis cultive sur hydrocarbures. Biochim. Biophys. Acta. 220: 373-385.

54. Walker, J. D., J. J. Cooney. 1973. Pathway of n-alkane oxidation in Cladosporium resinae. J. Bacteriol. 115: 635-639.

55. Lode, E. T., M. J. Coon. 1973. Role of rubredoxin in fatty acid and hydrocarbon hydroxylation reactions. In Iron-sulfur proteins (W. Lovenberg, ed.). Academic Press, New York. pp. 173-191.

56. Kawamoto, S., M. Ueda, C. Nozaki, M. Yamamura, A. Tanaka, S. Fukui. Localization of carnitine acetyltransferase in peroxisomes and in mitochondria of n-alkane-grown Candida tropicalis. FEBS Letters 96: 37-40.

57. Mishina, M., T. Kamiryo, S. Tashiro, S. Numa. 1978. Separation and characterization of two long-chain acyl-CoA synthetase from Candida lipolytica. Eur. J. Biochem. 82: 347-354.

58. Yamada, T., H. Nawa, S. Kawamoto, A. Tanaka, S. Fukui. 1980. Subcellular localization of long-chain alcohol dehydrogenase and aldehyde-dehydrogenase in n-alkane-grown Candida tropicalis. Arch. Microbiol. 128: 145-151.

59. Graves, L. B., Jr., W. M. Becker. 1974. Beta-oxidation in glyoxysomes from Euglena. Protozool. 21: 771-774.

60. Woodward, J., M. J. Merrett. 1975. Induction potential for glyoxylate cycle enzymes during the cell cycle of Euglena gracilis. Eur. J. Biochem. 55: 555-559.

61. Kobr, M. J., F. Vanderhaeghe, G. Combepine. 1973. Particulate enzymes of the glyoxylate cycle in Neurospora crassa. Biochem. Biophys. Res. Comm. 37: 460-645.

62. Szabo, A., C. J. Avers. 1969. Some aspects of regulation of peroxisomes and mitochondria in yeast. Ann. N. Y. Acad. Sci. 168: 302-312.

63. Sullivan, J. O., P. J. Casselton. 1972. The subcellular localization of glyoxylate cycle enzymes in Coprinus lagopus (sensu Buller). J. Gen. Microbiol. 75: 333-337.

64. Graves, L. B. Jr., V. N. Armentrout, D. P. Maxwell. 1976. Distribution of glyoxylate-cycle enzymes between microbodies and mitochondria in Aspergillus tamarii. Planta 132: 143-148.

65. Graves, L. B. Jr., R. N. Trelease, A. Grill, W. M. Becker. 1972. Localization of glyoxylate cycle enzymes in glyoxysomes in Euglena. J. Protozool. 19: 527-532.

66. Gerhardt, B. 1971. Localization of microbodial enzymes in Polytomella caeca. Arch. Mikrobiol. 80: 205-218.

67. Cooper, R. A., D. Lloyd. 1972. Subcellular fractionation of the colourless alga Polytomella caeca by differential and zonal centrifugation. J. Gen. Microbiol. 72: 59-70.

68. Tolbert, N. E. 1974. In Algal Physiology and Biochemistry (W. D. P. Stewart, eds.). Blackwell, Oxford. pp. 474-504.

69. Nelson, E. B., N. E. Tolbert. 1970. Glycolate dehydrogenase in green algae. Arch. Biochem. Biophys. 141: 102-110.

70. Frederick, S. E., P. Gruber, N. E. Tolbert. 1973. The occurrence of glycolate dehydrogenase and glycolate oxidase in green plants. An evolutionary survey. Plant Physiol. 52: 318-323.

71. Stabeau, H. 1976. Microbodies from Spirogyra. Organelles of a filamentous algae similar to leaf peroxisomes. Plant Physiol. 58: 693-695.

72. Codd, G. A., G. H. Schmid. 1972. Enzymic evidence for peroxisomes in a mutant of Chlorella vulgaris. Arch. Mikrobiol. 81: 264-272.

73. Gruber, P. J., S. E. Frederick. 1977. Cytochemical localization of glycolate oxidase in microbodies of Klebsormidium. Planta 135: 45-49.

74. Collins, N., M. J. Merrett. 1975. The localization of glycollate-pathway enzymes in Euglena. Biochem. J. 148: 321-328.

75. Collins, N., M. J. Merrett. 1975. Microbody marker-enzymes during transition from phototrophic to organotrophic growth in Euglena. Plant Physiol. 55: 1018-1022.

76. Yokota, A., Y. Nakano, S. Kitaoka. 1978. Different
    effects of some growing condition on glycolate
    dehydrogenase in mitochondria and microbodies in
    Euglena gracilis. Agric. Biol. Chem. 42: 115-120.
77. Yokota, A., Y. Nakano, S. Kitaoka. 1978. Metabolism
    of glycolate in mitochondria of Euglena gracilis.
    Agric. Biol. Chem. 42: 121-129.
78. Graves. L. B. Jr., R. N. Trelease, A. Grill, W. M.
    Becker. 1972. Localization of glyoxylate cycle
    enzymes in glyoxysomes in Euglena. J. Protozool.
    19: 527-532.
79. Paul, J. S., C. W. Sullivan, B. E. Volcani. 1975.
    Photorespiration in diatoms. Mitochondrial
    glycolate dehydrogenase in Cylindrotheca fusiformis
    and Nitzschia alba. Arch. Biochem. Biophys.
    169: 152-159.
80. Stabenau, H. 1974. Verteilung von Microbody-Enzyme
    aus Chlamydomonas in Dichtegradienten. Planta
    118: 35-42.
81. Colby, J., H. Dalton, R. Whittenbury. 1979. Bio-
    logical and biochemical aspects of microbial growth
    on C_1 compounds. Annu. Rev. Microbiol 33: 481-517.
82. Sahm, H., F. Wagner. 1973. Microbial assimilation of
    methanol. The ethanol- and methanol-oxidizing
    enzymes of the yeast Candida boidinii. Eur. J.
    Biochem. 36: 250-256.
83. Roggenkamp, R., H. Sahm, W. Hinkelmann, F. Wagner.
    1975. Alcohol oxidase and catalase in peroxisomes
    of methanol-grown Candida boidinii. Eur. J. Biochem.
    59: 231-236.
84. Veenhuis, M., I. Keizer, W. Harder. 1979. Characteri-
    zation of peroxisomes in glucose-grown Hansenula
    polymorpha and their development after the transfer
    of cells into methanol-containing media. Arch.
    Microbiol 120: 167-175.
85. Fukui, S., S. Kawamoto, S. Yasuhara, A. Tanaka. 1975.
    Microbody of methanol-grown yeasts. Localization
    of catalase and flavin-dependent alcohol oxidase
    in the isolated microbody. Eur. J. Biochem.
    59: 561-566.
86. Veenhuis, M., J. P. Van Dijken, W. Harder. 1976.
    Cytochemical studies on the localization of methanol
    oxidase and other oxidases in peroxisomes of
    methanol-grown Hansenula polymorpha. Arch.
    Microbiol. 111: 123-135.

87. Tanaka, A., S. Yasuhara, S. Kawamoto, S. Fukui, M. Osumi. 1976. Development of microbodies in the yeast Kloeckera growing on methanol. J. Bacteriol. 126: 919-927.

88. Bateman, D. F., S. V. Beer. 1965. Simultaneous production and synergistic action of oxalic acid and polygalacturonase during pathogenesis by Sclerotium rolfsii. Phytopathology 55: 204-211.

89. Maxwell, D. P., D. F. Bateman. 1968. Oxalic acid biosynthesis by Sclerotium rolfsii. Phytopathology 58: 1635-1642.

90. Armentrout, V. N., L. B. Graves, Jr., D. P. Maxwell. 1978. Localization of enzymes of oxalate biosynthesis in microbodies of Sclerotium rolfsii. Phytopathology 68: 1597-1599.

91. Chang, C. C., A. H. C. Huang. 1981. Metabolism of glycolate in isolated spinach leaf peroxisomes. Kinetics of glyoxylate, oxalate, CO_2 and glycine formation. Plant Physiol. 67: 1003-1006.

92. van Dijken, J. P., P. Bos. 1981. Utilization of amines by yeast. Arch. Microbiol. 128: 320-324.

93. Yamada, H., O. Adachi, K. Ogata. 1965. Amine oxidases of microorganisms. Part III. Properties of amine oxidase of Aspergillus niger. Agric. Biol. Chem. 29: 864-869.

94. Zwart, K., M. Veenhuis, J. P. Dijken, W. Harder. 1980. Development of amine oxidase-containing peroxisomes in yeast during growth on glucose in the presence of methylamine as the sole source of nitrogen. Arch. Microbiol. 126: 117-126.

Chapter Four

THE ROLE OF MICROTUBULES IN PLANT CELL WALL GROWTH

MYRON C. LEDBETTER

Biology Department
Brookhaven National Laboratory
Upton, NY 11973

INTRODUCTION

The earliest microscopic evidence that higher plants are composed of cells rested on visualization of walls as cell remnants in wood and cork (Figure 1). In time it was recognized that the living eukaryotic plant cell is usually encased in a wall which grows by apposition from within, and that crystalline cellulose predominates as a skeletal component of the wall. Cell shape and function are determined largely by the extent and pattern of wall material deposition, along with the orientation in which cellulosic elements are laid down. The longstanding interest in the physical and biological properties of walls is reflected in several summaries which have appeared on the subject through the years.[1-3]

Cellulose, in particular, has been intensively studied, as well it might be, considering that it represents the most abundant natural product of the biosphere, with a turnover rate estimated at 10^{11} tons per year.[4] The unit in which cellulose appears in the wall is the microfibril, a flattened structure about 10 to 15 nm wide and 4 to 7.5 nm thick in higher plants, running many micrometers in length. Though there is controversy over higher and lower orders of organization of cellulose,[1,2,5] it is generally accepted that the

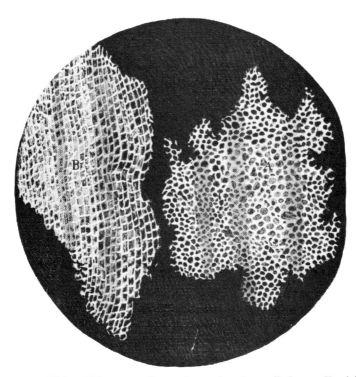

Figure 1. This illustration of cork, from Robert Hook's "Micrographia" of 1665 is believed to be the first published representation of plant cell walls.

first identifiable cellulosic product of the cell is synthesized at the plasma membrane and organized into a microfibril.

It was found from studies using polarized light, and confirmed by electron microscopy, that the way cellulose microfibrils are deposited in the wall varies from a highly oriented configuration in some cases, to a random array in others. The amount of parallelism and the direction of microfibrillar orientation relative to the main cell axis are characteristic of cell type and age, and of the particular wall examined. Microfibrils have a relatively high tensile strength, and their configuration is reflected in the physical nature of the wall. During the cell expansion stage of growth, the ability of the wall to stretch more in

one direction than another is determined, in part, by the
orientation of the microfibrils. This in turn, contributes
to the shape of the mature cell. The extraordinary strength
of some fibers, such as those of hemp and flax, is attribu-
table to the tensile properties of the microfibrils and their
orientation nearly parallel to the long axis of the cell.

A role for the cytoplasm in wall formation was recognized in the mid 1800's by Cruger[6] and Dipple.[7] These
investigators, working independently, examined cells
destined to develop secondary bands of thickening along
their walls. They found that bands of granular, streaming
cytoplasm overlay the bands of secondary thickening during
their development (Figure 2). More recent studies

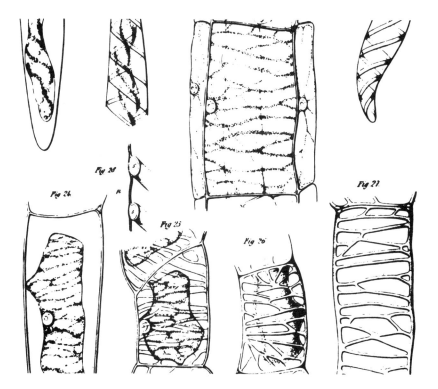

Figure 2. Drawings from Dipple's work,[7] which described
bands of streaming, granular cytoplasm in relation to the
developing bands of secondary wall thickening.

concerning the fine structure of cellulose biogenesis have
centered around the plasma membrane and the cortical
cytoplasm. The plasma membrane is of interest because it is
the immediate site of microfibril synthesis. It has been
found to contain arrays of particles that are believed to be
the synthase complexes responsible for generating the cellulose microfibrils. The cortical cytoplasm adjacent to the
plasma membrane has been examined for structures capable of
regulating and guiding the cellulose synthesis. Special
attention has been given to cortical microtubules as anisotropic elements of the cytoskeleton which possibly guide the
synthase complexes as they deposit new microfibrils upon the
wall.

THE CELL SURFACE AND THE LIMITING MEMBRANE

The fine structure of the cell surface and its limiting
membrane has been studied since the 1960's for evidence of
the mechanism which generates cellulose microfibrils.
Sectioning methods of that time revealed the membrane as a
relatively smooth tripartite entity lacking any particulate
or anisotropic structures, and appressed to the cell wall.
It has been assumed that the cellulose synthase mechanism
is composed of particles, and that these key elements are
at or within the plasma membrane. The search for such
particles, based on metal shadowed carbon replicas of
variously prepared materials, has led to two differing concepts of the structure of the mechanism and how it works.

Early studies, which led to the ordered granule hypothesis, were done by R. D. Preston and his colleagues on
the wall and membrane architecture of some algae and higher
plants. These workers used shadowed replicas of cell walls
which occasionally had pieces of the protoplast adhering to
them. The microfibrils of the wall were beautifully dis-
played in these replicas as mostly parallel ribbons,
deposited in layers (Figure 3). From one layer to the next
the microfibrils alternated in orientation. Adhering bits
of the protoplast, not removed during specimen preparation,
were sometimes observed to have 30 nm granules arranged in
a cubic lattice (Figure 4). An axis of the lattice was
reported to coincide with the microfibril direction, and
from such evidence, Preston hypothesized[8] that the granules
are the glucan synthases which extend the chains of the
microfibril as its end progresses along a row of granules

Figure 3. Metal shadowed carbon replica of inner wall of
Chetomorpha melagonium showing two microfibrillar orienta-
tions.[2]

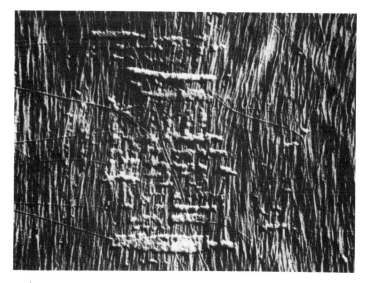

Figure 4. Micrograph similar to Fig. 3, showing portion of
wall to which some granular cytoplasm adhered following
plasmolysis.[2] Granules seen in this way formed the basis
for the ordered granule hypothesis.

within the lattice (Figure 5). He assumed that the granules
coat the outside of the plasma membrane, and that their
arrangment accounts for the three predominant directions of
microfibril orientation in walls, with synthesis proceeding
along the two major axes of particles and to a lesser
extent the diagonal. Particles of similar size have been
seen as hexagonal arrays in freeze-etch preparations of
yeast,[9] and in section.[10,11] This has been taken as supporting evidence for the ordered granule hypothesis; however, the granules are seen only occasionally in section
and, when they do appear, seem inappropriately disposed
with respect to both spacing and proximity to the plasma
membrane.

Some recent freeze-fracture studies of both higher
plants and algae at the laboratories of R. M. Brown[12] and
L. A Staehelin[5] have led to a concept quite different from
ordered granules. The specimens were prepared by quick-
freeze techniques which completely avoid the use of cryopro-
tectants. This is an important departure since spatial
distribution of intramembrane particles can be altered
during preparation by other methods. It has been shown,
for example, that upon plasmolysis membrane particles can
aggregate into close-packed arrays.[13] It is also important
in these studies to maintain turgidity of the protoplast so

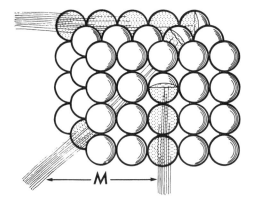

Figure 5. Representation of ordered cellulose synthase
granules in a cubic array of 30 nm particles. According to
the ordered granule hypothesis the granules form microfibrils (M) of cellulose by addition to the glucan chains
along the two major axes and the diagonal.[8]

that the membrane is tightly appressed to the wall. With the membranes fixed in this conformation, the cellulose microfibril is revealed as a ridge or groove, depending on which aspect of the membrane is viewed. The use of 20% glycerine as a cryoprotectant just before freezing was found to reduce turgidity enough to prevent the needed intimate membrane-wall contact.[12]

The organisms examined by these new techniques are phylogenetically quite different, one being an alga and the other a higher plant, yet there is an important constant in the results (Figures 6, 7). It appears that a rosette of intramembrane particles probably six in number, is associated with the end of a cellulose microfibril. The complex is some 24 to 28 nm in diameter and is thought to be composed of cellulose synthesizing enzymes. The complex is assumed to move within the plane of the membrane as the microfibril is spun out. There are differences in some details, such as the association of complexes with one another into close packed arrays in some instances, and the exact arrangement of the particles within the membrane (Figs. 6, 7); however, the rosette of synthase particles appears as a common characteristic. The detection of rosettes within cytoplasmic vesicles[5] supports the idea that the complexes are Golgi derived and are transported to the plasmalemma by vesicle fusion.

These two concepts of how the cellulose synthase system works differ in important respects. In the ordered lattice hypothesis it is assumed that there is a more-or-less fixed array of synthase particles on the exterior of the protoplast, and the microfibril grows by additions to glucan chains by successive particles within a row. In other words, it is the growing end of the microfibril which is supposed to move over a fixed array of particles. In the more recent concept, it is the rosette of particles which moves within the plane of the plasma membrane as the micro-fibril is generated. This difference is important in considering the possible role of the microtubules in determining the direction of microfibril synthesis.

It may be possible, in retrospect, to interpret the earlier reports of ordered arrays of particles from what is now known about the mobility of intramembrane particles. It seems probable that the earlier preparative procedures

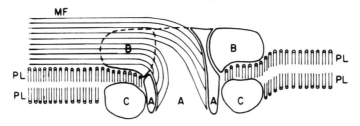

Figure 6. Model of cellulose synthase complex interpreted
from quick-freeze fracture images from higher plants,[12] with
the microfibril (MF), rosette of particles (C), and terminal
complex (A,B) in bimolecular phospholipid leaflets of plasma
membrane (PL).

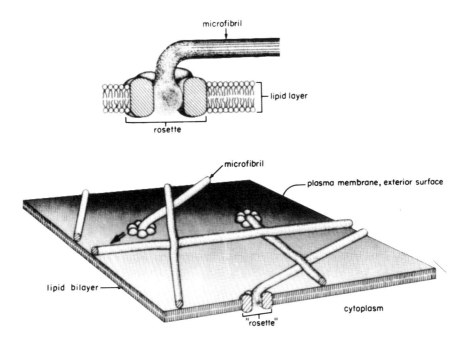

Figure 7. Model of microfibril formation in primary walls
of the alga _Micrasterias_ as interpreted from quick-freeze
fracture images.[5]

allowed the complexes, interpreted as granules, to become
crowded into rectangular or hexagonal arrays prior to
replication.[13] In contrast, the evidence indicates that
the newer rapid freezing methods arrest the particles in
place, where they can be detected at the ends of the
developing microfibrils.[4,12]

THE CORTICAL CYTOPLASM

Studies have been made of the cortical cytoplasm to
discover in it some anisotropic element(s) related to the
production of cellulose microfibrils and their orientation
during synthesis. These studies parallel those of the
plasmalemma and, like them, have been limited by the
available techniques, notably a number of chemical fixation
protocols for preparing specimens for thin sectioning. The
early potassium permanganate fixative gave way to osmium
tetroxide and then to glutaraldehyde-osmium tetroxide. The
structural components of interest in these studies have
been the endoplasmic reticulum, a clear zone or gap near
the plasma membrane, and the cortical microtubules and
their associated proteins.

Potassium permanganate, for a time the fixative of
choice for plant material, dramatically revealed membranes
as lines of high contrast but showed little else in the
cytoplasm (Figure 8). Using this fixative, Porter and
Machado studied dividing cells in onion root tips and noted
that rapidly growing walls were associated with a rich
display of the endoplasmic reticulum near the wall.[14] This
relationship, which has been confirmed by others, is not
understood and may be an indirect one.[15]

Veronal acetate buffered osmium tetroxide was in vogue
as a fixative for animal studies when permanganate was
favored by botanists. It was obvious that images from the
osmium fixed cells offered a richer and presumably more
faithful preservation of fine structure than did pre-
manganate, however, the results were quite unsatisfactory
when osmium was applied to plant material in veronal
acetate buffer. We had reason to suspect the difficulty
lay with the buffer, and we replaced the veronal acetate
with phosphate, finding from the literature that phosphates
are the principal natural buffers of plant cells.[16] The
results were rewarding with a wide variety of cell types.

We began to search the cell cortex for evidence of structures we could relate to wall formation; however, at the
place of most interest, just within the plasmalemma, we
frequently observed a clear zone or gap (Figure 9). Other
cytoplasmic features seemed well preserved, which led us to
discount the possibility that this gap was an artifact of
our fixation method.

Shortly after this we heard from Russell Barrnett's
laboratory that a student, David Sabatini, had achieved some
interesting results while looking for improved fixation for
cytochemical use. During his search Sabatini found that
cells treated with dialdehydes, particularly glutaraldehyde,
before osmium fixation showed improved morphological preservation, presumably because the dialdehydes act as cross-
linking agents for proteins and polyhydroxy compounds.[17]
When we applied this procedure to plant cells, we noticed
that the cytoplasm was more evenly filled with structures
up to the plasma membrane (Figure 10), including the
previous clear zone. Examination of the cell cortex
revealed slender, rather straight to gently curved elements
when caught in longitudinal section (Figure 11) and dense
rings with clear lumen when seen in cross sections (Figure
12). These microtubules[18] were about 24 nm in diameter and
could be seen to run for micrometers in length in favorable
sections. We realized that the microtubules lay parallel
to the known orientation of cellulose microfibrils in the
side walls of these barrel-shaped cells of root meristems
(Figure 13), and suspected they might have something to do
with microfibril orientation. This promoted us to search

Figure 8. Electron micrograph of thin section of meristematic higher plant cell fixed in potassium permanganate.
Some conspicuous membranes are the nuclear envelope (NE),
endoplasmic reticulum (ER), and plasma membrane (PM). The
primary cell wall (CW) appears as a low density zone
between adjacent cells.

Figure 9. Image of a cell similar to that in Fig. 8, but
fixed in phosphate buffered osmium tetroxide. The cell
wall (CW) is dense and is bounded by the plasma membrane
(PM). The cortical cytoplasm often shows clear zones or
gaps (arrows).

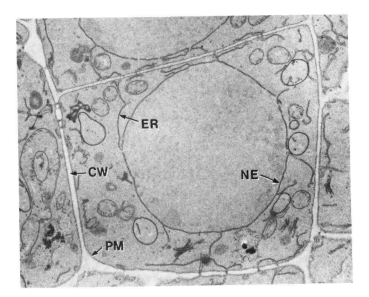

Figure 8.

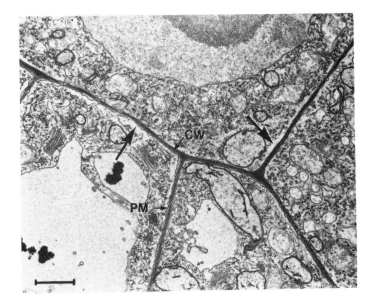

Figure 9.

for grazing sections of end walls, where the cellulose was
known to be randomly arranged (Figure 14). At these end
walls the microtubules again mirrored in orientation the
recently deposited cellulose microfibrils (Figure 15). This
coincidence of orientation between cortical microtubules and
adjacent cellulose microfibrils has since been noted in many
cell types and has been discussed in several reviews of
microtubules.[19-22] In the cortex of cells undergoing
primary wall growth the microtubules are more-or-less
evenly dispersed (Figure 15). In banded secondary thickening the microtubules are confined to the thickening part
of the wall (Figures 16 and 17). Some exceptions to the
rule of parallelism in orientation between microtubules
and adjacent microfibrils have been noted[21] and one of
these, which we investigated, is briefly discussed below.

An often quoted exception to this parallelism is that
found by Preston and Goodman[23] in a study of new wall
formation in Cladophora repestris swarmers. This alga
discharges naked swarmers, or zoospores, which begin to
synthesize a new wall shortly after the cells are freed.
Preston and Goodman reported microtubules to be present in
the naked swarmers but absent at any stage when there was
a detectable wall. Lawrence Crockett and I reinvestigated
this alga using Cladophora gracilis, the species available
at Woods Hole, Massachusetts.[24] Zoospores were fixed in
buffered glutaraldehyde at frequent, short intervals following their release as naked cells. Approximately 15
minutes after discharge, under our conditions, the cells
had developed a thin but discernable wall. The cortical
cytoplasm contained microtubules (Figure 18), a finding
contrary to that of Preston and Goodman.[23] In a later

Figure 10.   Portion of a meristematic higher plant cell
fixed in glutaraldehyde prior to osmium tetroxide. Structures are preserved evenly in the cell cortex without the
clear zones noted in Fig. 9.

Figure 11.   Small portion of cortex and wall of higher
plant cell fixed in glutaraldehyde-osmium tetroxide,
showing microtubules (Mt) adjacent to the plasma membrane
(PM). Microtubules lie circumferentially about the
principal axis of this cell cut in transverse section.

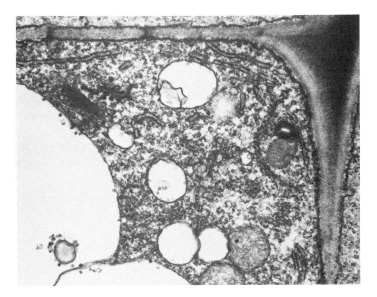

Figure 10.

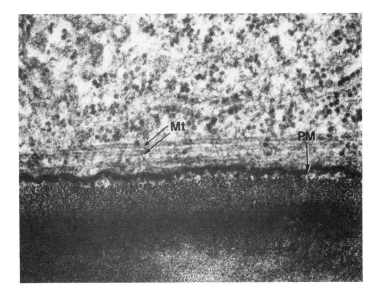

Figure 11.

report from Preston's laboratory[25] their earlier findings
with Cladophora are confirmed and details of their methods
are given which may offer a clue to the reason for the difference between their results and ours. They held the cells
at 8°C and then fixed them at 3°C, whereas we did our fixation at room temperature of about 20°C. Microtubules are
known to depolymerize at low temperature,[19] and it seems
likely that the lower temperature may have permitted depoly-
merization before the microtubules were fixed, especially if
the newly formed wall delays penetration of the fixative to
the cortical cytoplasm. There were other differences between
their investigation and ours, including the use of different
species; nevertheless, it seems reasonable to conclude from
the results of our work with the fixative at a higher
temperature that the relationship between cortical microtubules and cellulose microfibrils in Cladophora is, in
important respects, similar to that reported for most other
wall forming plant cells that have been investigated.

MICROTUBULE-PLASMALEMMA RELATIONSHIPS

The available evidence suggests that the cortical
microtubules in plant cells influence the direction in which
the cellulose synthase complexes move during microfibril
generation, though the mechanism by which this is done is
unknown. Heath has proposed that cellulose synthase complexes may be guided through the fluid matrix of the
plasmalemma by means of links to the microtubules, which
then function as tracks for directional movement of the
complexes.[26] Indeed, suitable bridges have been
reported[10,18,27] though they have not been quantified.

Figure 12. Portion of cell similar to that in Fig. 11 but
cut in longitudinal section with the plasma membrane (PM)
and microtubule (Mt) cut normal to the plane of section.

Figure 13. Section longitudinal to the principal cell axis
and almost parallel to the side wall separating two cells
to the right and left, showing the cell wall (CW), plasma
membrane (PM), and cell cortex (CC) viewed nearly en face.
Parallel arrays of microtubules populate the cortical
cytoplasm. The dense structures seen in groups are plasmodesmata organized into pit fields.

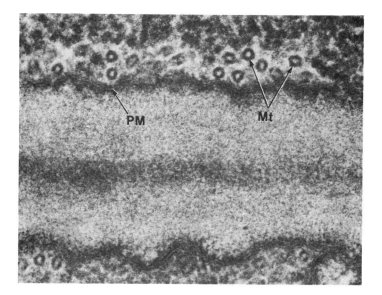

Figure 12.

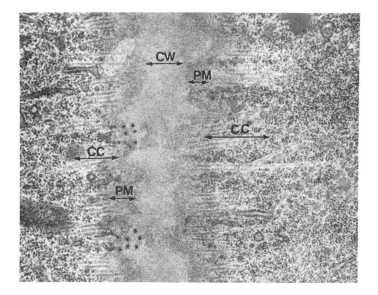

Figure 13.

Charles E. Anderson and I have begun a study of the relationship between microtubules and the plasmalemma using flax fibers (<u>Linum</u> <u>usitatissimum</u>). Some of our preliminary impressions are briefly discussed here. We have chosen to study fibers engaged in S-2 thickening of the cell walls.[28] The fibers are easily identifiable (Figure 19) as single or groups of thick-walled cells within the stem chlorenchyma. The S-2 phase in these fibers can be characterized as the production of an extensive thickening of the wall (Figure 20) in which the microfibrils are oriented almost axially[1,3] as we have confirmed by SEM of the inner wall surface (Figure 21). Bridges have been noted between the microtubules and the membrane (Figure 22). The cortical microtubules usually appear as groups of several more-or-less evenly spaced elements.[18] We have followed several microtubules through serial sections for as much as one micrometer in length, with some showing bridges in every section observed and few entirely free from the membrane for that length. The general impression is that the typical cortical microtubule in these fiber cells is frequently attached to the plasma membrane by bridges which are spaced such that each section of about 50 to 60 nm in thickness may contain a bridge. Furthermore, a rather constant minimal distance of about 7 nm separates the microtubule and the membrane, suggesting the existence between these structures of a spacer of constant length as might be provided by the bridges. Presumably the bridges are composed of microtubule associated proteins (MAPs).[29]

Figure 14. Grazing section including portions of the end wall (CW), and the cortical cytoplasm with its microtubules (Mt). The microtubules, like the adjacent microfibrils of cellulose, are disposed at random.

Figure 15. Diagram of an interphase meristematic cell showing the spherical nucleus in the center, and microtubules in the cell cortex adjacent to the plasma membrane. Broken lines in the wall represent the orientation of cellulose microfibrils randomly arranged in the end and parallel in side walls. In both the end and side walls the microtubules mirror in orientation the adjacent microfibrils.

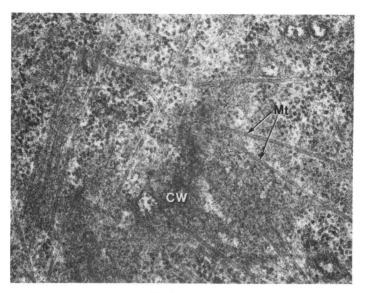

Figure 14.

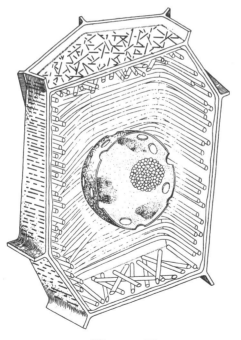

Figure 15.

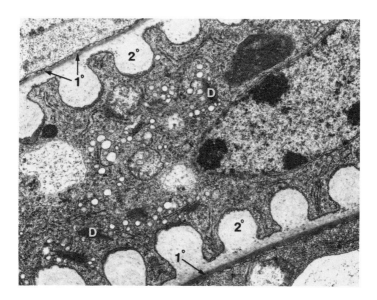

Figure 16. Near median longitudinal section of a dif-
ferentiating xylem cell in a primary leaf of Phaseolus
vulgaris. Secondary bands of thickening (2°) in rings or
spirals are being deposited on the primary wall (1°). The
cytoplasm contains numerous dictyosomes (D) and vesicles.

Figure 17. Secondary thickening showing the plasma membrane
(PM) and microtubules (arrows), some of which show attach-
ment bridges to the plasma membrane. The microtubules are
confined to the cortex over the thickening band and are
absent between thickenings. Both microtubules and microfibrils of the band lie normal to the plane of the section.

Figure 18. Portion of a zoospore of Cladophora gracilis
fixed about 15 minutes following release. A thin cell wall
(CW) and cortical microtubules (arrows) are present. The
microtubules do not radiate from the nearby basal body (B)
since their axes run nearly parallel to one another.

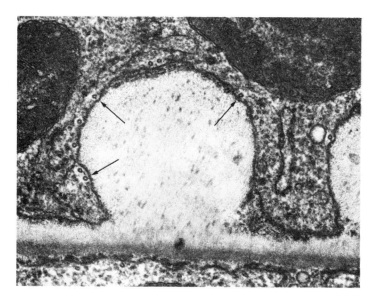

Figure 17.

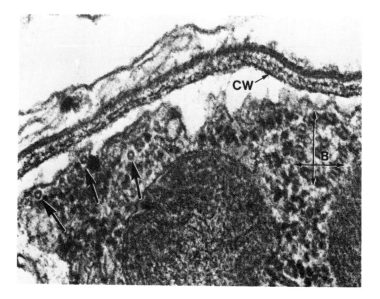

Figure 18.

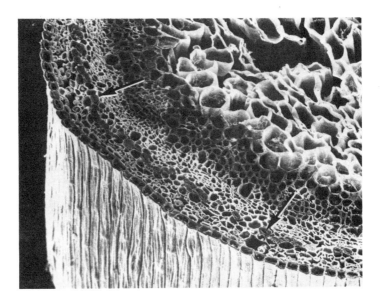

Figure 19. Scanning electron micrograph of a portion of
flax stem showing a transversely cut surface. Large pith
cells, some disrupted, are to the upper right and the epidermis is to the lower left. A discontinuous band of thick-
walled fiber cells appears within the stem chlorenchyma
about four to six cell layers within the epidermis (arrows).
Vascular tissue is found between the fibers and the pith.

     The spatial relationship between the microtubules, the
bridges, and the cellulose synthase complexes remains unknown, and it will surely be needed for an understanding
of how the system works. Current knowledge presents little
impediment to Heath's suggestion that the complexes are
guided by means of links to the microtubules. There may be
other explanations; for instance, it seems just as likely
that microtubule-plasmalemma bridges establish linear zones
of exclusion between which the complexes are free to move
in a linear fashion, or possibly the microtubules move in
relation to the wall and carry along attached complexes.
Whatever the mechanism, it is clear that microtubules are
not required to propel the complexes through the fluid
matrix of the membrane, since randomly oriented depositions

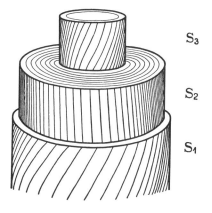

Figure 20. Diagram of <u>Linum</u> fiber cell showing the relationship of the three secondary thickenings.[3] Line on the
outer surface of the thickening indicate the predominant
orientation of cellulose microfibrils. The extensive S-2
thickening is of particular interest.

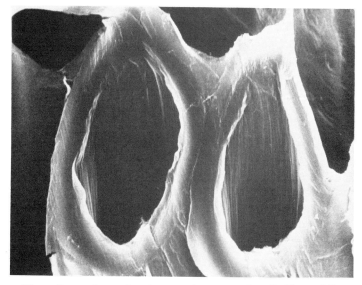

Figure 21. Scanning electron micrograph of flax fibers cut
about 45° to the principal cell axis. The cell contents
were removed with deoxycholate to expose the inner cell wall.
Vertical striations on the inner surface of the newly formed
S-2 wall confirm the nearly axial orientation of recently
deposited microfibrils.

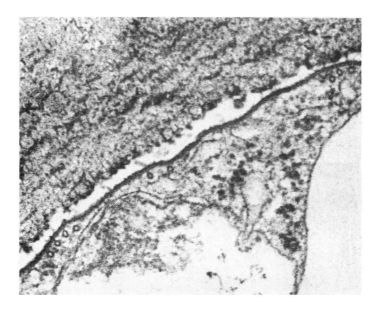

Figure 22.   Portion of flax fiber cut in cross section with
a group of cortical microtubules adjacent to the S-2 wall.
The microtubules are separated from the membrane by about
7 nm, with some microtubules showing bridges to the plasma
membrane.

occur in the absence of microtubules in root hair tips[30]
and in the presence of colchicine,[31] which disrupts microtubules.   There is evidence that an intact plasma membrane
is required for active cellulose synthesis.[32]

CONCLUSIONS

The microtubule remains the most likely candidate for
regulatory controller of oriented cellulose deposition
during wall growth.   It is clear that cortical microtubules
are not required for cellulose synthesis, though when they
are present the orientation of the microfibrils parallels
that of the microtubules.   At least in some systems, the
microtubules are linked to the plasmalemma by frequent
bridges, probably MAPs.   The plasmalemma contains cellulose
synthase complexes composed of intramembrane particles

shown to be arranged into rosettes in some algae and higher plants. These complexes must move through the fluid matrix of the plasmalemma as they spin out cellulose which becomes organized into microfibrils. The morphological relationships of the microtubules, the bridges, and the synthase complexes are unknown. Insight into these relationships would help us to understand how the system works.

Other questions that still need answering concern the regulation of such factors as microtubule orientation and organization,[33,34] biosynthesis of the cellulose,[35] rates and localization of synthesis, etc. Much remains to be learned about the synthesis of the most abundant natural product of the biosphere.

ACKNOWLEGEMENTS

The early work on the cortical cytoplasm discussed here was begun while I was associated with Keith R. Porter in his laboratory at Rockefeller University, and continued with Porter at the Biological Laboratories, Harvard University. Later work at Brookhaven National Laboratory has been done with the collaboration of L. C. Crockett, City College and P. S. Woods, Queens College, New York, and Charles A. Anderson, North Carolina State University, Raleigh, and with many valuable discussions on plant morphogenesis with Wm. J. Crotty of New York University. Portions of the work were aided at various times by the technical assistance of Carolyn Trager Burr, Walter Geisbusch, Richard Ruffing, and Ruth Wright. Gratitude is expressed for these associations as well as for agency support which has come from the U.S. Public Health Service, National Science Foundation, U.S. Department of Agriculture, and U.S. Department of Energy and its predecessors.

REFERENCES

1. Frey-Wyssling, A. 1976. The plant cell wall. Borntraeger, Berlin, 294 p.
2. Preston, R. D. 1974. The physical biology of plant cell walls. Chapman and Hall, London, 491 p.
3. Roelofsen, P. A. 1949. The plant cell wall. Borntraeger, Berlin, 335 p.

4. Hess, K. 1928. Die Chemie der Zellulose und ihrer
    Begleiter. Akademische Verlagsgesellschaft M.B.H.,
    Leipzig.
5. Giddings, T. H., Jr., D. L. Brower, L. A. Staehelin.
    1980. Visualization of particle complexes in the
    plasma membrane of Micrasterias denticulata
    associated with the formation of cellulose fibrils
    and secondary cell walls. J. Cell. Biol. 84: 327-339.
6. Cruger, H. 1855. Zur Entwicklungsgeschichte der
    Zellwand. Bot. Ztg. 13: 601-613, 617-629.
7. Dipple, L. 1867. Die Entstehung der wandstandigen
    Protoplasmastromchen. Abb. Naturforsch. Ges.
    Halle 10: 53-68.
8. Preston, R. D. 1964. Structural plant polysac-
    charides. Endeavour 23: 153-159.
9. Moor, H., K. Muhlethaler. 1963. Fine structure in
    frozen-etched yeast cells. J. Cell Biol.
    17: 609-628.
10. Robards, A. W. 1969. Particles associated with
    developing plant cell walls. Planta 88: 376-379.
11. Roland, J.-C. 1967. Aspects infrastructuraux des
    relations existant entre le protoplasme et la paroi
    des cellules de collenchyme. J. Microscopie
    9: 399-412.
12. Mueller, S. C., R. M. Brown, Jr. 1980. Evidence for
    an intramembrane component associated with a cellu-
    lose microfibril-synthesizing complex in higher
    plants. J. Cell Biol. 84: 315-326.
13. Wilkinson, M. J., D. H. Northcote. 1980. Plasma
    membrane ultrastructure during plant protoplast
    plasmolysis, isolation and wall regeneration: a
    freeze-fracture study. J. Cell Sci. 42: 401-415.
14. Porter, K. R., R. D. Machado. 1960. Studies on the
    endoplasmic reticulum IV. Its form and distribu-
    tion during mitosis in cells of onion root tip. J.
    Biophys. Biochem. Cytol. 7:167-180.
15. Crispeels, M. J. 1980. The endoplasmic reticulum.
    In The plant cell (N. E. Tolbert, ed). The bio-
    chemistry of plants (P. K. Stumpf, E. E. Conn, eds.)
    Vol. I. Academic Press, New York, pp. 389-412.
16. Miller, E. C. 1938. Plant physiology. Edit. 2.
    McGraw-Hill, New York, 1201 p.

17. Sabatini, D. D., K. Bensch, R. J. Barrnett. 1963.
    Cytochemistry and electron microscopy. The preser-
    vation of cellular ultrastructure and enzymatic
    activity by aldehyde fixation. J. Cell Biol.
    17: 19-58.
18. Ledbetter, M. C., K. R. Porter. 1963. A "microtubule"
    in plant cell fine structure. J. Cell Biol.
    19: 239-250.
19. Dustin, P. 1978. Microtubules. Springer-Verlag,
    Berlin, 452 p.
20. Hepler, P. K. Plant microtubules. In Plant Bio-
    chemistry, 3rd edit. (J. Bonner, J. E. Varner,
    eds.). Academic Press, New York, pp. 147-187.
21. Hepler, P. K., B. A. Palevitz. 1974. Microtubules
    and microfilaments. Annu. Rev. Plant Physiol.
    25: 309-362.
22. Newcomb, E. H. 1969. Plant microtubules. Annu. Rev.
    Plant Physiol. 20: 253-288.
23. Preston, R. D., R. N. Goodman. 1968. Structural
    aspects of cellulose biosynthesis. J. Roy. Micro-
    scopical Soc. 88: 513-527.
24. Crockett, L. J., M. C. Ledbetter. 1970. The associa-
    tion of microtubules with early wall formation in
    the zoospores of the marine alga Cladophora
    gracilis. (Abstract) Amer. J. Bot. 57: 741.
25. Robinson, D. G., R. K. White, R. D. Preston. 1972.
    Fine structure of swarmers of Cladophora and
    Chaetomorpha III. Wall synthesis and development.
    Planta 107: 131-144.
26. Heath, I. B. 1974. A unified hypothesis for the role
    of membrane bound enzyme complexes and microtubules
    in plant cell wall synthesis. J. Theor. Biol.
    48: 445-449.
27. Cronshaw, J. 1967. Tracheid differentiation in
    tobacco pith cultures. Planta 72: 78-90.
28. Cote, W. 1977. Wood ultrastructure in relation to
    chemical composition. In The structure, biosyn-
    thesis, and degradation of wood (F. A. Loewus,
    V. C. Runeckles, eds.). Plenum, New York, pp. 1-44.
29. Sloboda, R. D. 1980. The role of microtubules in
    cell structure and cell division. Amer. Scientist
    68: 290-298.
30. Bonnett, H. T., Jr., E. H. Newcomb. 1966. Coated
    vesicles and other cytoplasmic components of growing
    root hairs of radish. Protoplasma 62: 59-75.

31. Pickett-Heaps, J. D. 1967. The effects of colchicine on the ultrastructure of dividing plant cells, xylem wall differentiation, and distribution of cytoplasmic microtubules. Dev. Biol. 15: 206-236.

32. Carpita, N. C., D. P. Delmer. 1980. Protection of cellulose synthesis in detached cotton fibers by polyethylene glycol. Plant Physiol. 66: 911-916.

33. Gunning, B. E. S., A. R. Hardham, J. E. Hughes. 1978. Evidence for initiation of microtubules in discrete regions of the cell cortex in Azolla root-tip cells, and an hypothesis on the development of cortical arrays of microtubules. Planta 143: 161-179.

34. Gunning, B. E. S. 1979. Nature and development of microtubule arrays in cells of higher plants. Proc. 37th Annual Electron Microscopy Society of America Meeting. pp. 172-175.

35. Delmer, D. P. 1977. The biosynthesis of cellulose and other plant cell wall polysaccharides. In The structure, biosynthesis and degradation of wood (F. A. Loewus, V. C. Runeckles, eds.). Plenum, New York. pp. 45-77.

Chapter Five

PHOTOSYNTHETIC CARBON METABOLISM IN CHLOROPLASTS

STEVEN C. HUBER

U.S. Department of Agriculture, Science and
Education Administration, Agricultural Research
Service and Departments of Crop Science and
Botany, N.C. State University
Raleigh, North Carolina   27650

INTRODUCTION

The purpose of this review is to discuss some aspects
of current research interest dealing with carbon metabolism
in mesophyll chloroplasts of C_3 plants that may be involved
in control of the rate of carbon fixation or the distribution of fixed carbon among products. These aspects are
stressed because an appreciation of the metabolic regulation
of chloroplasts is necessary in order to begin to understand
the coordinated metabolism that occurs between the chloroplast and the cytosol in situ. Specifically, this review
will emphasize recent developments indicating the important
role of stromal pH on component processes which regulate the
photosynthetic rate and utilization of inorganic phosphate
[P_i], and also the mechanisms which may be involved in the
control of stromal pH.

Important aspects that will not be covered here but
are discussed elsewhere include photosynthesis in C_4 plants
(chapter by Campbell and Black) and Crassulacean Acid Metabolism plants;[1] electron transport and photophosphorylation;[2] and properties of the chloroplast envelope including
metabolite transport.[3-5] Excellent general reviews of the

biochemistry of chloroplasts have been provided recently by
Halliwell[6] and Jensen.[7]

The enzymes of the Calvin cycle, which function in the
autocatalytic fixation of carbon dioxide, are contained
within the chloroplast stroma, as are the enzymes for starch
formation. In the dark, starch is mobilized and the glucose
metabolized by a glycolytic sequence or by the oxidative
pentose phosphate pathway.[8,9] Enzymes involved in the formation of sucrose, dicarboxylic acids, and certain amino
acids, however, are localized in the cytosol (Figure 1).

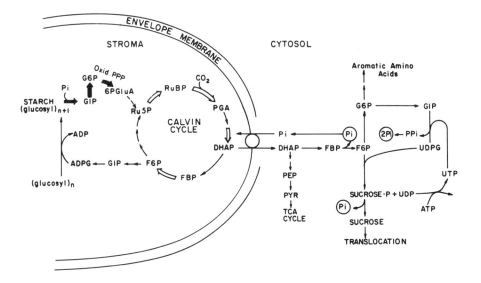

Figure 1. Simplified scheme showing the compartmentation of
reactions involved in CO_2 assimilation leading to the formation of carbohydrates (starch and sucrose), certain amino
acids, and substrates for mitochondrial respiration. RuBP,
ribulose 1,6-P_2; PGA, 3-P glycerate; DHAP, dihydroxyacetone-
P; F6P, fructose-6-P; FBP, fructose-1,6-P_2; G1P, glucose-
1-P; G6P, glucose-6-P; ADPG, ADP-glucose; UDPG, UDP-glucose;
PEP, phosphoenolpyruvate; Pyr, pyruvate; 6-PG1uA, 6-P-
gluconate, Oxid. PPP, oxidative pentose phosphate pathway.
Circled P_i indicates sites of P_i release during sucrose
formation. The circle in the membrane is the phosphate
translocator. Open and closed arrows indicate enzymes that
are activated and inactivated by light, respectively.

Movement of carbon skeletons across the chloroplast envelope
is thought to involve a specific carrier protein in the
inner membrane of the envelope. The transport system,
referred to as the phosphate translocator, releases a molecule of triose phosphate in exchange for the uptake of a
molecule of P_i.[10,11] Hence, overall cellular metabolism
involves the coordinated functioning of the chloroplast
(which fixes CO_2 and consumes P_i) and the cytosol (which
releases P_i) and a phosphate balance is maintained on both
sides of the membrane. The behavior of isolated chloroplasts, described below, reflects the division of labor
depicted in a highly simplified form in Figure 1.

REGULATION BY INORGANIC PHOSPHATE $[P_i]$

When chloroplasts are isolated with intact outer
envelope membranes, the stromal proteins and cofactors
necessary to catalyze CO_2 fixation are retained and high
rates of photosynthesis can be achieved in vitro (50 to 150
μmol O_2 evolved/mg chlorophyll·h). The principal end-
products of CO_2 assimilation are dependent upon experimental
conditions but usually include starch, triose phosphates,
and glycolate, an intermediate of the photorespiratory
cycle. Under conditions which reduce photorespiratory glycolate formation (e.g., high CO_2, low O_2, and low pH),[11a]
starch and triose phosphates are the primary products, and
their relative formation is controlled by the concentration
of exogenous inorganic phosphate $[P_i]$.[12,13]

Photosynthesis can be monitored polarographically as
CO_2-dependent O_2 evolution. The advantage of monitoring O_2
evolution (as opposed to $^{14}CO_2$ assimilation) is that contin-
uous traces are obtained. The typical responses obtained
with wheat chloroplasts, as affected by increasing concentrations of P_i (Figure 2 inset), are similar to those originally obtained with spinach chloroplasts (for review see
reference 14). Because the chloroplast is a P_i-consuming
organelle,[15] the photosynthetic rate is low and non-linear
with time in the absence of external P_i (for sake of
clarity, the minus P_i curve is not shown in the inset of
Figure 2). With increasing concentrations of P_i in the
medium, rates of carbon fixation (and O_2 evolution) increase
to a maximum and then decline (Figure 2). In general, as
the concentration of P_i is increased there is a progressive
increase in the time required for the onset of

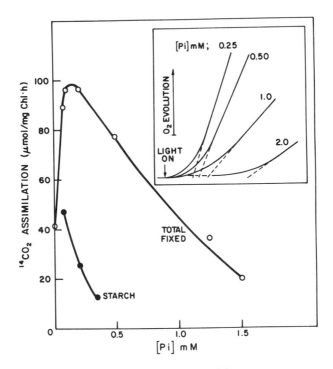

Figure 2.   Phosphate dependence of $^{14}CO_2$ assimilation and
^{14}C-starch formation by intact wheat chloroplasts.   The inset shows the effect of P_i on the kinetics of CO_2-dependent
O_2 evolution.   The reaction mixture contained 0.34 M
sorbitol, 50 mM HEPES-NaOH (pH 7.60), 1 mM $MgCl_2$, 1 mM
$MnCl_2$, 2 mM EDTA, 7 mM $NaH^{14}CO_3$ (1 μCi/μmol), 200 units/ml
catalase, P_i as indicated, and chloroplasts (20 μg Chl/ml).
Chloroplasts were isolated as described in reference 22.

photosynthesis.   The lag phase, or induction period, represents the time required for chloroplast metabolite pools
and enzyme activities to reach levels sufficient to sustain
maximum turn-over of the cycle.   Inhibition of photosynthesis at high P_i concentrations may be caused by excessive
depletion of internal metabolites as a result of enforced
export by P_i.[14]   Another factor may be reduced activity of
certain Calvin cycle enzymes (see discussion below).

Maximum starch formation generally occurs at a lower P_i concentration than that of the fixation rate (Figure 2 and references 12 and 13). As the concentration of P_i in the medium is increased, triose phosphate release from the chloroplast is enhanced and fixed carbon is diverted from starch formation. Starch formation is reduced primarily because of a decreased P-glycerate/P_i ratio in the stroma, which inhibits ADP glucose-pyrophosphorylase.[13,16] Reduced labeling of starch (by $^{14}CO_2$ assimilation) may result from inhibition of formation as well as increased starch degradation. Net labeling of starch should reflect the balance between the two processes.

The P_i dependence of wheat chloroplast photosynthesis (Figure 2) typifies the response originally characterized with spinach chloroplasts. Qualitatively similar responses have been observed with chloroplasts from bundle sheath cells of the C_4 plant Panicum miliaceum,[17] the CAM plant Sedum praealtum,[18] peas,[19] tobacco (Huber, unpublished) and rhodoplasts of the red algae Griffithsia monilis.[20] With CAM chloroplasts, however, the inhibition of photosynthesis at superoptimal P_i concentrations was less pronounced than is typically observed (e.g., Figure 2) and the induction phase was not significantly increased.[18] Hence, C_3 chloroplasts all appear to contain a phosphate translocator, but characteristics may vary depending on the species.

The phosphate translocator has been studied most extensively in spinach chloroplasts. The transport catalyzed appears to be a strict exchange reaction of the influx of one molecule in exchange for the efflux of another. Certain characteristics of the phosphate translocator of spinach chloroplasts are summarized in Table 1. Transport of a compound on the phosphate translocator can be studied either as inhibition of $^{32}P-P_i$ uptake or by the release of $^{32}P-P_i$ from preloaded chloroplasts (back exchange technique). P_i, 3-P-glycerate [PGA], dihydroxyacetone-P [DHAP], and glyceraldehyde-3-P [GAP] are readily transported, as evidenced by high affinities (i.e., low K_m's) for the translocator and high V_{max} (Table 1). In contrast, compounds such as pyrophosphate [PP_i], citrate and G6P are transported at an almost insignificant rate (low V_{max}). However, PP_i and citrate appear to interact with the carrier, as evidenced by relatively low K_i's for inhibition of P_i uptake. Inhibition results because binding to the carrier renders it

Table 1.   Kinetic constants for transport on the phosphate
translocator of spinach chloroplasts.[a]

Compound	$^{32}P-P_i$-uptake K_i (mM)	$^{32}P-P_i$ back exchange K_m (mM)	V_{max}
P_i	[$K_m(P_i)$ = 0.20 mM	0.30	57
PGA	0.15	0.14	36
DHAP	0.13	0.13	51
GAP	--	0.08	41
PP_i	1.8	--	<1
Citrate	1.5	--	<1
G6P	40	--	<1

[a]Data from reference 11.

immobile, whereas PGA, DHAP, and GAP inhibit P_i uptake as
a result of their simultaneous transport. Action of the
translocator can also be inhibited by the sulfhydryl reagent
p-chloromercuriphenyl sulfonate and pyridoxal phosphate,
which indicates the requirement of a free sulfhydryl group
and lysine residue for operation of the carrier.[11]

     Factors that affect the utilization of exogenous P_i
during chloroplast photosynthesis in vitro are of interest
from the standpoint of understanding the behavior of the
isolated organelle and also as they may relate to interactions with the cytoplasm in situ. The concentration of
P_i at which photosynthetic rates attain maximum values
(e.g., about 0.15 mM in Figure 2) represents the balance
point where P_i uptake matches the production of triose phosphates for export, such that transport neither limits nor
exceeds fixation. The point at which balance is achieved
depends upon the kinetic constants of the phosphate translocator and the concentration of transport metabolites in the

stroma and medium (and their rate of production). Because
of the high degree of specificity of the phosphate translocator, the concentrations of P_i, PGA and DHAP will be the
most important determinants of transport. P_i curves for
photosynthesis (e.g., Figure 2) can be shifted to higher or
lower concentrations depending upon experimental conditions.
The effect of a number of experimental parameters are
grouped according to the process probably affected, and the
effects on P_i utilization are summarized in Table 2.
Certain seemingly trivial factors such as ratio of Chl to
reaction volume and light versus dark pretreatment of leaf
tissue have large effects, and become important when comparing results among different experiments. Preillumina-
tion of leaves prior to chloroplast isolation apparently
increases stromal metabolite pools. Hence, the induction
phase (at a given P_i concentration) is reduced and the P_i
optimum is increased. Similar effects are produced by an
increase in the Chl/reaction volume ratio. The latter
observation indicates that accumulation of metabolites in
the medium is as important as increased levels in the stroma.
This point is emphasized further by the observation that
inclusion of alkaline phosphatase in the reaction medium
dramatically lengthens induction and inhibits photosynthesis
(Figure 3). Presumably, alkaline phosphatase increases the
concentration of external P_i as DHAP is converted to the
nontransport metabolite dihydroxyacetone.

Given constant experimental conditions of the type
discussed above (i.e., pre-illumination status of leaves and
a fixed Chl/volume ratio), utilization of P_i is generally
affected by changes in either transport or CO_2 fixation.
Inhibition of P_i uptake by addition of a chemical modifier
(pCMBS or pyridoxal-P) or nontransported metabolite
inhibitor (citrate or PP_i), decreases the induction phase,
and correspondingly increases the P_i optimum (Table 2).

Factors that affect CO_2 fixation can be divided into
two groups: (a) those that reduce cycle turnover and,
hence, cause an obligatory decrease in V_{max}, and (b) those
that affect cycle activity without an obligatory reduction
in V_{max} (Table 2). Maximum photosynthetic rate can be
limited by reduction of energy supply (i.e., low light or
high light plus an electron transport inhibitor such as
DCMU). When photosynthesis is energy-limited, the induction
phase and P_i optimum are unaltered. This indicates that

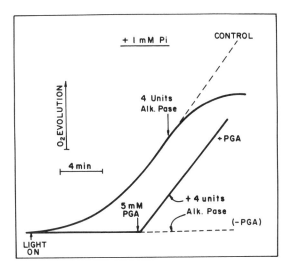

Figure 3. Inhibition of spinach chloroplast O_2 evolution by alkaline phosphatase (2 units/ml) and reversal by PGA. Chloroplasts were isolated as described in reference 29, and assay conditions were as described in legend of Figure 2 (Huber, unpublished).

relatively low cycle activity is required to fill the metabolite pools necessary to terminate induction. However, when turnover of the cycle is inhibited by glyceraldehyde,[30] the P_i optimum is reduced and the induction phase is increased. These effects can not be explained solely by reduced CO_2 fixation. Rather, they probably relate to the specific mode of action of glyceraldehyde, which is thought to inhibit by acting as an alternate substrate in the transketolase reaction.[30a] As such, the effect of glyceraldehyde would be to reduce stromal metabolite levels. Hence, glyceraldehyde should shift the P_i optimum to lower concentrations. Qualitatively similar effects are produced by reduced external pH or reduced osmotic support in the reaction medium (Table 2). However, these conditions do not necessarily reduce maximum photosynthetic rates. The effects of pH will be discussed in more detail below, but both factors may increase the pool size of stromal metabolites necessary to sustain maximum photosynthetic rates.

Table 2. Factors that affect the utilization of P_i during chloroplast photosynthesis.

| Parameter affected | Example | Effect on O$_2$ evolution | | | Ref. |
		P_i optimum	V_{max}	Induction Phase	
Increased stromal metabolites	Pre-illumination of leaves	Increased	No effect	Decreased	14,21
Increased metabolites in medium	Increased Chl/volume	Increased	No effect	Decreased	22
Inhibition of P_i uptake	PP$_i$	Increased	No effect	Decreased	19,23
	Pyridoxal-P or pCMBS	Increased	No effect	Decreased	24,25
Reduced Calvin cycle activity (Type a)	Low light	No effect	Decreased	No effect	26,27
	Electron transport inh.	No effect	Decreased	No effect	26,27
	Glyceraldehyde	Decreased	Decreased	Increased	24,27
Increased stromal metabolite levels required for photosynthesis (Type b)	Low external pH	Decreased	No effect or decreased	Increased	27,28
	Low osmotic support	Decreased	No effect	Increased	27

Reduced osmotic support would induce chloroplast swelling
and thus increase stromal volume.[31,32] The concentration
of metabolites necessary to terminate induction would be
unaltered, but because of increased volume, the necessary
pool size (in terms of μg-atoms of carbon) would be increased. Consequently, the induction phase would increase
and the P_i optimum would decrease, without a significant
effect on maximum rate at the P_i optimum (provided that
osmolarity is not reduced to the point that chloroplast
integrity is not maintained).

    Considering all of the responses above, two generali-
zations emerge. First, factors that increase the induction
phase invariably reduce the P_i optimum and second, the P_i
dependence of photosynthesis is not directly related to the
rate of carbon flux through the cycle. Increased metabolite levels (both in the stroma and external medium) necessary for termination of induction are probably derived from
CO_2 assimilation as well as from hexose metabolism (derived
from starch) by glycolysis or the oxidative pentose phosphate pathway.[33]

INDUCTION PHENOMENA AND ENZYME ACTIVATION

    Following a dark to light transition, an induction
period is typically observed for photosynthetic CO_2 uptake
(or O_2 evolution) not only with isolated chloroplasts but
also with isolated leaf cells and intact leaves.[34] With
isolated chloroplasts, considerable evidence suggests that
termination of the induction period requires increased
levels of cycle intermediates (for review see reference 14).
Recent evidence with isolated chloroplasts suggests that
enzyme activation may also be involved, and may be the
principal factor with more organized systems, such as
mesophyll protoplasts and leaves.[35]

    Chloroplasts contain a number of enzymes that are modu-
lated by light (for review see reference 36). Enzymes of
the Calvin cycle are activated by light whereas enzymes involved in starch mobilization and hexose degradation are
inactivated by light[37] (open and closed arrows, respectively,
in Figure 1). Light-activated enzymes include NADP-malate
dehydrogenase, NADP-GAP dehydrogenase, ribulose-5-P [Ru5P]
kinase, fructose bisphosphatase [FBPase] and sedoheptulose
bisphosphatase [SBPase]. Depending upon the enzyme, and to

some extent the species from which it is obtained, illumination causes a two- to eight-fold increase in activity of
carbon fixation enzymes.

At present, the mechanism responsible for light modulation in situ is not entirely clear. Enzyme activation is
thought to occur either as a result of disulfide reduction[38]
or intramolecular disulfide exchange reaction,[39] which then
result in increased enzyme activity. Reducing equivalents
are supplied by electron transport, or in vitro by the
reductant dithiothreitol [DTT],[40] but the mechanism of
transfer to the enzymes is controversial. Anderson and
Avron[41] have postulated involvement of thylakoid bound
"light effect mediators" (LEM) which directly transfer
reducing equivalents to the soluble stromal enzymes. In
contrast, Wolosiuk and Buchanan[42] have postulated that electrons are transferred from ferredoxin, via soluble thioredoxin to the soluble enzymes. Several forms of thioredoxin
have been identified with specificity for the enzymes
activated.[43] For FBPase a third mechanism has also been
identified. A new iron-sulfur protein ("ferralterin") has
been isolated that catalyzes a light-dependent activation of
FBPase in the presence of only thylakoid membranes.[44] It
is not known whether other enzymes are also activated by
this mechanism. In addition, the relative importance of
the presently recognized mechanisms to enzyme activation in
situ is also unclear.

One critical question concerning enzyme activation is
whether the kinetics are sufficiently slow to contribute to
the induction phase of photosynthesis. In a recent study
with wheat chloroplasts, Leegood and Walker[45] observed a
seven-fold activation of FBPase and smaller activations of
NADP-GAP dehydrogenase and Ru5P kinase prior to the onset
of O_2 evolution. Addition of PGA or DHAP to the reaction
mixture, however, significantly shortened the induction
period without affecting the kinetics of enzyme activation.
They concluded, therefore, that induction can be accounted
for solely in terms of increased metabolite concentrations.
Charles and Halliwell[46] also concluded that the rate of
FBPase activation by light was not rate-limiting for CO_2
fixation either during or after the induction phase.

In contrast, other studies have concluded that the
induction period is dependent, at least partially, upon

activation of Calvin cycle enzymes. Heldt et al.[35,47]
reported that illumination of spinach leaf protoplasts
(which exhibited an initial induction phase of about 1 min)
or intact spinach leaves, did not result in large changes
in metabolite levels. Hence, other factors may be involved.
With isolated chloroplasts several Calvin cycle enzymes
were activated by light with the half time for activation
ranging from 0.5 min for Ru5P kinase to 2.8 min for SBPase.
In studies from our laboratory,[48] the activation of Ru5P
kinase and GAP-dehydrogenase occurred over several minutes
in intact barley chloroplasts and importantly, the kinetics
of activation were affected by conditions which altered the
length of the induction phase. Exogenous P_i lengthened the
lag and slowed activation, whereas PGA reduced the lag and
accelerated activation. Furthermore, in a reconstituted
system of thylakoid membranes with concentrated stromal
proteins, enzyme activation by light was shown to be sensitive to the substrate/P_i ratio.[48] More recently, Furbank
and Lilley[49] showed that high P_i inhibits Calvin cycle
turnover in a reconstituted system free of thylakoid membranes. Inhibition occurred at the level of sedoheptulose
and fructose bisphosphatase. Hence, direct effects of P_i
upon the Calvin cycle are significant. Subsequently, Heldt
et al.[47] also have demonstrated that activation of FBPase
and SBPase in intact spinach chloroplasts was inhibited by
P_i and accelerated by PGA. Overall, the results suggest
that the enzyme activation mechanism(s) may be regulated by
stromal metabolite levels. It is not clear why metabolite
levels affect enzyme modulation by light in some, but not
all studies.

        The basis for metabolite effects on enzyme modulation
by light remains to be elucidated. Recently, however, a
requirement for substrate in the activiation of FBPase[50] and
SBPase[51] has been identified. Activation of purified FBPase
by DTT and rate-limiting amounts of thioredoxin required
FBP.[50] The effect of FBP was abolished when the concentration of thioredoxin was saturating. In the case of SBPase,
activation of the partially purified enzyme by DTT absolutely required Mg^{2+} + SBP.[51] Thioredoxin was not added to
the reaction mixtures, but the SBPase preparation probably
contained a low concentration of thioredoxin as a contaminant. Some activation of SBPase can be obtained with DTT
in the absence of thioredoxin, but complete activation
requires DTT plus thioredoxin.[52] Interestingly, in these

studies, a requirement for SBP during activation was not detected.[52] It appears then that the apparent dependence of activation on substrate may vary depending upon the source and purity of the enzyme and perhaps the concentration of thioredoxin present during activation. Further study is required to determine whether there are fundamental differences between FBPase and SBPase in terms of substrate requirements for activation. Nevertheless, enzyme modulation by light in intact chloroplasts, whatever the mechanism, responds to changes in the levels of P_i and organic phosphates (substrates). Identification of the basis for metabolite effects may help to resolve the mechanism of enzyme modulation that occurs in situ.

Metabolite levels may exert a general control on enzyme modulation by light. It was reported that the inactivation by light of G6P dehydrogenase in intact chloroplasts was prevented by concentrations of P_i that inhibited photosynthesis[33] (Figure 4). The effect of P_i on both photosynthesis and G6PDH activity was reversed upon addition of the Calvin cycle intermediate PGA (Figure 4). At lower concentrations of P_i, the kinetics of enzyme inactivation paralleled the kinetics of O_2 evolution. It was postulated that

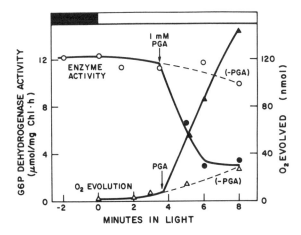

Figure 4. Kinetics of O_2 evolution and inactivation of G6P dehydrogenase by light in barley chloroplasts in the presence of 1 mM P_i. At arrow, 1 mM PGA was added to the reaction mixtures. Adapted from reference 33.

during the induction phase, phosphorolytic break-down of starch[12,53] and hexose metabolism via the oxidative pentose phosphate pathway may be involved in the formation of cycle intermediates required for the termination of the lag.[33] At present, the relation between enzyme modulation and metabolite levels is not entirely clear. The picture that emerges is that changes in metabolite levels may influence the activation state of several key enzymes. Changes in enzyme activity may affect the onset of photosynthesis after a dark period (induction) as well as the steady state rate of carbon fixation.

STROMAL pH AND CATION FLUXES

Upon illumination of intact chloroplasts, photoinduced electron transport results in the movement of protons from the stroma to the intrathylakoid space. As a result, the pH of the stroma increases from about pH 7.3 in the dark to pH 8.0 in the light.[54] There are several important consequences of proton pumping to photosynthesis. First, the pH gradient generated across the thylakoid membrane (about 2.5 pH units) provides the driving force for photophosphorylation.[55] Second, stromal pH affects both the rate of CO_2 assimilation by the Calvin cycle, as well as the utilization of exogenous P_i during CO_2 fixation (see Table 2).

Experimentally, the pH of the stroma of illuminated chloroplasts can be manipulated several different ways. For example, salts of weak acids (e.g., sodium formate) reduce the stromal pH by dissipating the pH gradient across the envelope[56,57] because the molecule can traverse the envelope both as the undissociated molecule (free acid) and as the anion.[58] Hence, protons can be transferred indirectly across the envelope. Stromal pH can also be reduced by lowering the pH of the reaction mixture,[56] which indicates that although the envelope is relatively impermeable to protons, some direct or indirect proton transfer can occur in the absence of exogenous mediators. For a decrease in external pH of one unit, the stromal pH is reduced about 0.6 pH units.[27,54] Significant proton permeability of the envelope is suggested by "pH-jump" experiments. When isolated chloroplasts are allowed to initiate photosynthesis in a weakly buffered suspension at an optimal external pH (e.g., pH 7.8), a rapid reduction in external pH (e.g., to pH 7.3) affects the rate of photosynthesis within 10 sec

and a new steady state rate is achieved in about 3 min
(Huber, unpublished). It can be concluded that the new
stromal pH level was attained within this relatively rapid
time span.

Regardless of the means by which stromal pH is reduced,
a reduction in the rate of CO_2 assimilation is observed.
Analysis of the products formed during CO_2 assimilation at
suboptimal pH revealed an overall decrease in total cycle
intermediates but increased levels of FBP and SBP.[11a, 57-59]
The results suggest that the inhibition of photosynthesis
caused by low stromal pH may be attributed to inhibition of
fructose- and sedoheptulose-bisphosphatases. In the case
of FBPase, low stromal pH may reduce catalytic activity by
decreasing the affinity of the enzyme for FBP [i.e.,
increased Km (FBP)]. In studies with the purified chloroplasts enzyme, Zimmerman et al.[60] reported that a decrease
in pH from 8.8 to 8.2 increased the Km (FBP) from 0.08 mM
to 0.45 mM. Hence, the concentration of FBP required to
maintain the maximal reaction velocity would be increased
dramatically as the pH is lowered. Also, the activation of
FBPase by light may be reduced under conditions of low
stromal pH which would further reduce catalytic activity.

The effects of exogenous Mg^{2+} on chloroplast reactions
may also be mediated by effects on stromal pH. It is recognized that millimolar concentrations of free Mg^{2+} inhibit
photosynthesis in intact chloroplasts isolated from a number
of species.[61-63] Inhibition by Mg^{2+} is surprising for
several reasons: (a) in situ the chloroplasts must function
in the cytosol which must contain free Mg^{2+}; (b) the chloroplast envelope is relatively impermeable to Mg^{2+} [64] yet
internal processes are affected; and (c) Mg^{2+} is not
generally regarded as a toxic substance [e.g., stromal
Mg^{2+} is absolutely required for operation of the Calvin
cycle].[65]

The effects of Mg^{2+} on photosynthesis can be explained
on the basis that Mg^{2+} reduces the stromal pH of illuminated
chloroplasts. It has been postulated that Mg^{2+} activates a
K^+/H^+ exchange across the envelope membrane.[63,66] The
direction of cation flux is determined primarily by the K^+
and H^+ gradients between the stroma and the medium. When
the concentration of K^+ in the medium is low, stromal K^+ is
exchanged for H^+ from the medium (Figure 5A). As a result,

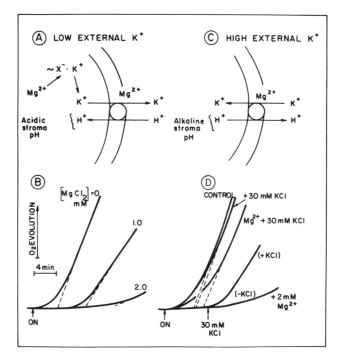

Figure 5.   Schematic representation of cation fluxes and typical polarograph traces obtained with spinach chloroplasts showing the effect of external K concentration on chloroplast processes in the presence of Mg^2 .   Adapted from reference 66.

stromal pH is reduced, which inhibits photosynthesis. As shown by the polargraph results in Figure 5B, in the absence of added Mg^{2+}, O_2 evolution exhibits the typical induction period before attaining a maximum rate. Addition of 1 or 2 mM Mg^{2+} to the chloroplast suspension prior to illumination increases the induction phase and causes substantial inhibition of the final rate. High concentrations of external K^+ prevents the release of stromal K^+, and associated influx of protons (Figure 5C). Hence, stromal pH is not reduced and photosynthetic rates are not impaired. Exogenous K^+ not only prevents the Mg^{2+}-induced stromal acidification but also reverses the effect, as evidenced by a rapid rise in O_2 evolution following addition of K^+ to Mg^{2+}-inhibited chloroplasts (Figure 5D).

Effects of Mg^{2+} on stromal pH are maximal within several minutes and occur when the Mg^{2+} is added to the chloroplast suspension in the light or in the dark.[63] However, Mg^{2+} is significantly less inhibitory to O_2 evolution when added in the light (after photosynthesis has started) as compared to the dark.[62] Hence, operation of the Calvin cycle apparently becomes less sensitive to changes in stromal pH after CO_2 assimilation has been initiated. This apparent anomaly can be rationalized on the basis that the response of photosynthesis to stromal pH is decreased as the concentration of P_i is decreased.[27] Assimilation of CO_2 results in increased pools of cycle intermediates and decreased P_i and, consequently photosynthesis can continue in the presence of Mg^{2+}.

Apparently most of the inhibition of O_2 evolution caused by Mg^{2+} can be attributed to reduced stromal pH.[67] Depletion of stromal K^+ may also contribute, to a lesser extent, to the observed inhibition. Conceivably, stromal K^+ concentration could affect the ratio of bound versus free-Mg^{2+} [68] or affect the transfer of electrons from photosystem I to O_2 during pseudocyclic electron flow.[69] Whether these or other processes are affected by changes in stromal K^+ remains to be established, as does the exact stoichiometry between fluxes of K^+ and H^+.

From the data in the literature, it may be adduced that Mg^{2+} binds to a "site" on the chloroplast envelope and in so doing, activates a reversible K^+/H^+ exchange. The nature of the putative Mg^{2+}-binding site remains to be established, but could be either a specific protein or the lipids of the envelope membrane. In general, little specificity has been observed for monovalent (K^+, Na^+, Cs^+, Rb^+) and divalent (Mg^{2+}, Mn^{2+}, Ca^{2+}) cations, insofar as they have been tested. The exceptions, however, are worth noting. In the case of monovalent cations, the organic cation choline was relatively ineffective in preventing Mg^{2+}-inhibition of photosynthesis.[66] Hence, the monovalent cation effect, when observed, is not simply an ionic strength displacement of Mg^{2+} from the envelope. Of the divalent cations, only Ba^{2+} was virtually inactive in causing inhibition of

chloroplast photosynthesis.[67] This high degree of specificity indicates that screening of membrane surface charges
cannot explain the effects of Mg^{2+}. In the latter case,
Ba^{2+} is not completely inert because Ba^{2+} antagonizes the
inhibitory action of Mg^{2+}.[70] Previously, Demmig and
Gimmler[63] reported Mg^{2+}/Ca^{2+} antagonism in the inhibition
of O_2 evolution. The understanding of the full significance of these experimental observations must await
further studies of the divalent cation binding site.

In the absence of Mg^{2+} the chloroplast envelope is
generally regarded as being relatively impermeable to both
H^+ and K^+. As a result, isolated intact chloroplasts
retain high concentrations of monovalent cations.[64] Some
cation transport across the envelope may be induced by
light. Gimmler et al.[64] reported that illumination of
intact spinach chloroplasts resulted in a transient efflux
of H^+ that was associated with a counter transport of K^+.
In recent studies, Kaiser et al.[68] reported that external
monovalent cations are essential for high rates of photosynthesis in isolated chloroplasts. In a "low-salt" assay
medium, CO_2-dependent O_2 evolution was inhibited almost
completely and could be restored to control rates by 50 to
100 mM K^+, Na^+ or RbCl. When deprived of external monovalent cations, the light-induced alkalization of the
stroma was found to be greatly reduced, as was the stromal
content of free-Mg^{2+}.[68] Low Mg^{2+} and reduced pH would be
expected to inhibit photosynthesis by decreased activity of
Calvin cycle enzymes. In continued studies by Gimmler and
coworkers,[71] it was postulated that stromal alkalization is
a prerequisite for K^+-uptake, rather than a consequence
thereof. The causal relation between stromal alkalization
and K^+ uptake was suggested on the basis that reduction of
the trans-envelope H^+ gradient (with a permeating weak acid)
inhibited K^+ uptake. It was postulated that in the light,
proton pumping into the thylakoid space is not completely
charge-compensated by the counter exchange of Mg^{2+} into the
stroma. The negative electrical potential then provides
the driving force for the light-dependent uptake of K^+ into
the stroma. Uptake of K^+ would allow greater H^+ pumping
into the thylakoid space (and thereby increase stromal pH)
and may also increase the free-Mg^{2+} concentration by competing for fixed cation binding sites in the stroma. Both
conditions would favor increased photosynthesis.

In terms of cation transport across the envelope, two important questions remain. First, whether transport occurs in the absence of divalent cations, and second, whether the fluxes are energy linked. A possible point of confusion concerns the inadvertent binding of divalent cations to the envelope during the isolation procedure. If such binding occurs, then the observed cation fluxes may be similar to the fluxes that occur in the presence of added Mg^{2+}. In studies conducted in our laboratory, light intensity during growth of spinach plants was found to greatly affect the apparent binding and retention of divalent cations to the envelope. Plants grown in high light (e.g., greenhouse) tended to retain bound divalent cations, whereas plants grown under lower light intensity (e.g., growth chamber or shaded greenhouse) did not. The response to external K^+ of CO_2-dependent O_2 evolution by chloroplasts isolated from spinach plants grown under high light is shown in Figure 6. The dramatic stimulation by K^+ of O_2 evolution shown in Figure 6A is associated with increased stromal pH (Huber, unpublished). Importantly, the stimulation by K^+ is prevented by 2 mM EDTA, which suggested that residual divalent cations bound to the envelope may be involved. Prewashing the chloroplasts with an EDTA-containing solution increased the rates of photosynthesis and eliminated the response to external K^+ (Figure 6B). Photosynthesis in chloroplasts from plants grown under low light intensities is unaffected by external K^+ in the absence of exogenous divalent cations (Figure 5D and reference 62). Hence, a dependence of chloroplast photosynthesis upon high concentrations of external monovalent cations is not always apparent. When a dependence is observed, it may be because of failure to remove divalent cations bound to the envelope.

Unfacilitated diffusion of cations across the envelope must certainly occur and such movement of protons could be extremely important. In order to maintain the H^+ gradient across the envelope, the passive H^+ influx from the medium (or cytosol) must be compensated by active H^+ extrusion. At present, there are two postulated mechanisms for energized H^+ extrusion. The first involves an ATP driven H^+ pump (or H^+/K^+ exchanger) at the envelope,[67] and the second envisions the electron transport chain as being responsible for initiating and maintaining an alkaline stroma[72] (Figure 7). The two mechanisms are fundamentally different, in

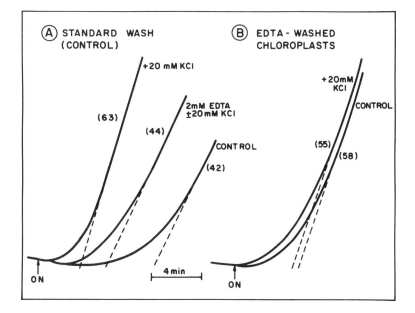

Figure 6.   Stimulation by external K^+ of CO_2-dependent O_2 evolution by spinach chloroplasts isolated from plants grown in the greenhouse. Chloroplasts were washed once with resuspension medium[29] without (A) or with (B) 2mM free EDTA. Assay conditions as in Figure 2 legend (Huber, unpublished). Maximum rates of O_2 evolution, expressed as μmol O_2 evolved/mg Chl·h, are shown in parentheses.

terms of the membrane across which protons are moved and whether ATP is involved.

Evidence in support of a role of the envelope ATPase in active H efflux involves the use of oligomycin, an ATPase inhibitor. The envelope ATPase is inhibitited by oligomycin (maximum inhibition of 40 to 70%) whereas the thylakoid ATPase (coupling factor) is completely insensitive.[67] Hence, oligomycin can be used to specifically probe the involvement of the envelope ATPase in photosynthesis. When supplied to intact chloroplasts that are actively assimilating CO_2, oligomycin was found to:[67]   (a) increase the inhibition of photosynthesis caused by acetate or Mg^{2+}; and

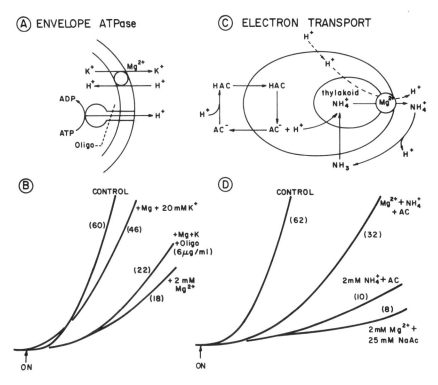

Figure 7.   Postulated mechanisms for energy linked proton efflux from the chloroplast stroma.   (A) and (B) adapted from references 66 and 67; scheme (C) from reference 72. Polarograph traces in (D) obtained in this laboratory are similar to those in reference 72.   Other conditions as in Figure 2 legend.   Maximum rates of O_2 evolution, expressed as µmol O_2 evolved/mg Chl·h, are shown in parentheses.

(b) prevent the reversal of Mg^{2+}-inhibition by low concentrations of external K^+ (Figure 7B).   These results suggested that the envelope ATPase may play a role in maintaining an alkaline stroma.   It was postulated[67] that H^+ fluxes across the envelope occur via an active, (oligomycin-sensitive) H^+ efflux and a reversible, Mg^{2+}-dependent (oligomycin-insensitive) H^+/K^+ exchange (Figure 7A).

The primary line of evidence to suggest that an ATP-driven H^+ flux is not involved comes from studies that

involved the energy transfer inhibitor Dio-9. Enser and
Heber[72] and Gimmler et al.[71] independently reported that
Dio-9, at concentrations that blocked the light-induced
increase in stromal ATP, had no effect on stromal alkalization. However, chloroplasts usually contain a significant
amount of ATP even in the dark[72,73] (about 10 to 14 nmol
ATP/mg Chl, which corresponds to 0.4 to 0.6 mM). The possibility that the dark level of ATP (presumably maintained
by substrate level phosphorylation) may be sufficient to
drive energized H^+ efflux cannot be ruled out. As an alternate mechanism, Enser and Heber[72] postulated that diffusion
of H^+ from the medium into the stroma may be removed (under
certain conditions) by the electron transport chain (Figure
7C). They made the interesting observation that restoration
of photosynthesis inhibited by formate (which discharges the
H gradient across the envelope) required $Mg^{2+} + NH_4^+$.[72]
Their observations have been confirmed with typical results
shown in Figure 7D. Enser and Heber[72] postulated that
electron transport drives the accumulation of NH_4^+ ions in
the intrathylakoid space, and that both H and NH_4^+ ions can
be released from the thylakoid space to the medium (facilitated in some way by Mg^{2+}). The net result is transport
of protons from the stroma to the medium. The significance
of this mechanism (Figure 7C) to the situation that exists
in situ, or in the absence of Mg^{2+} and NH_4^+, remains to be
established. A different interpretation of the results may
be that ammonia (NH_3) diffuses across the envelope. Formation of ammonium ions (NH_4^+) in the stroma would reduce the
concentration of protons and hence, increase stromal pH.
If ammonium ions were transported on the Mg^{2+}-dependent
cation exchanger (which remains to be established), export
of NH_4^+ could occur in exchange for external K^+ (Na^+). The
net result would be alkalization of the stroma and restoration of photosynthesis. Further work will be required to
distinguish between these two possible explanations.

CHLOROPLASTS AND CELLULAR METABOLISM

     Transport of metabolites across the chloroplast enve-
lope (catalyzed by the phosphate translocator) plays a
central role in cellular carbon metabolism (Figure 1).
Some of the most direct evidence comes from mannose-feeding
experiments.[74] Briefly, mannose is taken up by leaf tissue
and phosphorylated by cytoplasmic hexokinase to form
mannose-6-P, which in many species cannot be metabolized

further. As a result, cytoplasmic P_i is reduced. Under
these conditions, photosynthetic rates are reduced slightly,
but starch formation is dramatically increased.[75]

Additional evidence comes from studies with isolated
leaf protoplasts. Because protoplasts retain all necessary
enzymes, cofactors, etc. in the cytosol, the products of
$^{14}CO_2$ fixation by protoplasts are similar to those of
intact leaves. The distribution of labeled metabolites
(from $^{14}CO_2$ assimilation) between the chloroplast and
cytosol revealed that more than 90% of the fixed carbon was
outside the chloroplasts after a period of 10 min.[76] The
concentration of triose phosphates was 13-fold higher in the
cytosol than in the chloroplast, and it was concluded that
such uphill transport was driven by an opposite P_i gradient
catalyzed by the phosphate translocator.[76] Protoplasts
also exhibit an induction phase that can be altered by
changes in the external pH or osmolarity of the reaction
medium (Table 3), without necessarily affecting the final
rate of photosynthesis. Presumably, effects imposed on
internal pH or cellular volume are relatively transient in
nature. The effects of pH and osmolarity of the medium on
induction with isolated protoplasts (Table 3), are qualitatively similar to those obtained with isolated chloroplasts
(cf Table 2). Further, conditions that increased the induction phase (low pH or reduced osmolarity) increased
the partitioning of fixed carbon into sucrose (formed in
the cytosol) and decreased formation of starch (formed in
the chloroplast; Table 3). Altered partitioning patterns
are observed only after relatively short (10 min) exposures
to $^{14}CO_2$. After longer exposures, treatment differences
were not apparent which is consistent with the lack of
treatment effect on the final photosynthetic rate. The
results are entirely consistent with the correlation
between increased induction and greater export of cycle
intermediates (to the cytosol for sucrose formation), and,
hence, operation of the phosphate translocator in vivo.

The photosynthetic function of a fully expanded leaf
is to provide assimilates for the various metabolic "sink"
regions in the plant. In most higher plants, sucrose is
the principal translocatable end product of photosynthesis.
Transport of sucrose also continues in the dark (with carbon
supplied from starch mobilization). There is considerable
variation among species in photosynthetic formation

Table 3. Effects of pH and osmolarity of the reaction medium on induction and carbohydrate formation after 10 min of $^{14}CO_2$ fixation by wheat protoplasts.[a]

Variable	Induction phase (min)	$^{14}CO_2$ incorporation (% Total ^{14}C fixed)	
		Sucrose	Starch
pH 7.5			
0.35 M sorbitol	3.2	45	11
0.50 M sorbitol	2.1	43	12
0.70 M sorbitol	1.0	40	17
0.5 M Sorbitol			
pH 8.0	1.8	41	17
pH 7.5	2.1	44	14
pH 7.0	5.0	57	13

[a]Protoplasts isolated from 9-day old seedlings (Huber, unpublished).

of starch relative to sucrose. Partitioning of carbon between starch and sucrose appears to be biochemically controlled by factor(s) within the mesophyll cell.[77] The question then is whether product formation is controlled by the chloroplast or the cytosol. It appears that the rate of conversion of triose phosphates to sucrose (resulting in liberation of P$_i$.) may be the controlling factor. Recent evidence suggests that sucrose-P synthetase is the rate-limiting step in the sucrose formation pathway.[78] Variation among species in the activity of sucrose-P synthetase was correlated positively with the potential for sucrose formation and correlated negatively with leaf starch accumulation (Figure 8). Hence, starch and sucrose compete for fixed carbon and the control appears to be exerted in the cytosol,

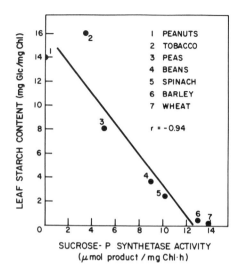

Figure 8. Comparison of leaf starch accumulation with activity of sucrose-P synthetase in cell-free extracts of leaves from various species.[78]

with the chloroplast in situ responding to changes in available P_i.

CONCLUDING REMARKS

Recent results suggest that the induction phase associated with the onset of chloroplast photosynthesis involves both increased levels of cycle intermediates and activation by light of certain enzymes. It is also apparent that enzyme activation may be regulated, in part, by metabolite levels. In isolated mesophyll protoplasts and intact leaves, where metabolite levels remain high and relatively constant during light-dark transitions, enzyme activation may be the primary factor responsible for termination of induction. Identification of the mechanism(s) responsible for enzyme modulation in situ and the control by metabolites, remain important areas for future research.

The pH of the chloroplast stroma is now recognized as an important factor involved in the regulation of photosynthesis. At suboptimal stromal pH, the catalytic potential of certain key enzymes such as FBPase is decreased as a result of reduced light activation and also because of a direct effect of pH on enzyme kinetic constants. With an increased Km (FBP) at low pH, the metabolite levels necessary to sustain the maximum photosynthetic rate are increased. Hence, the P_i optimum for chloroplast photosynthesis shifts to lower concentrations. In order to maintain an alkaline stromal pH, the passive influx of protons across the envelope must be accomodated by energized H efflux. Such proton movement may involve the envelope ATPase and/or electron transport chain.

The regulation of chloroplast processes by P_i reflects the coordinated metabolism that occurs between the chloroplast and cytosol in situ. Utilization of P_i during photosynthesis by isolated chloroplasts is affected by stromal pH, inhibitors of the phosphate translocator (certain P-esters), and to some extent, by the rate of carbon fixation. The variation in partitioning of fixed carbon between starch and sucrose, known to exist among species, may be related to the rate of conversion of triose phosphates to sucrose in the cytosol. Hence, chloroplast processes in vivo may be controlled by the cytosol, which regulates the availability of P_i.

ACKNOWLEDGEMENTS

Cooperative investigations of the North Carolina Agricultural Research Service and the Science and Education, Agricultural Research Service, United States Department of Agriculture, Raleigh, North Carolina. I am grateful to Drs. G. E. Edward, J. M. Robinson and A. R. Portis, Jr., for reading the manuscript and making helpful suggestions.

REFERENCES

1. Kluge, M. 1979. The flow of carbon in crassulacean acid metabolism (CAM). In Photosynthesis II (M. Gibbs, E. Latzko, eds.). Encyclopedia of Plant Physiology, New Series, Vol. 6. Springer-Verlag, New York. pp. 113-125.
2. Trebst, A. 1974. Energy conservation in photosynthetic electron transport of chloroplasts. Annu. Rev. Plant Physiol. 25: 423-358.

3. Douce, R., J. Joyard. 1979. Structure and function of the plastid envelope. Adv. Bot. Res. 7: 2-116.

4. Heber, U., H. W. Heldt. 1981. The chloroplast envelope: structure, function, and role in leaf metabolism. Annu. Rev. Plant Physiol. 32: 139-168.

5. Walker, D. A. 1974. Chloroplast and cell-The movement of certain key substances, etc. across the chloroplast envelope. In Plant biochemistry (D. H. Northcote, ed.). MTP Int. Rev. Sci. Biochem. Ser. I, Vol. 11. Butterworths, London. pp. 1-49.

6. Halliwell, B. 1978. The chloroplast at work. A review of modern developments in our understanding of chloroplast metabolism. Prog. Biophys. Molec. Biol. 33: 1-54.

7. Jensen, R. G. 1980. Biochemistry of the chloroplast. In The plant cell (N. E. Tolbert, ed.). The Biochemistry of Plants, Vol. 1, Chap. 7. Academic Press, New York. pp. 274-314.

8. Stitt, M., T. ap Rees. 1980. Carbohydrate breakdown by chloroplasts of Pisum sativum. Biochim. Biophys. Acta 627: 131-143.

9. Furbank, R. T., R. McC. Lilley. 1981. Reductive pentose phosphate cycle and oxidative carbohydrate metabolic activities in pea chloroplast stroma extracts. Plant Physiol. 67: 1036-1041.

10. Heldt, H. W. 1976. Metabolite transport in intact spinach chloroplasts. In The intact chloroplast (J. Barber, ed.). Elsevier/North-Holland Biomedical Press, The Netherlands. pp. 215-234.

11. Fliege, R., U. Flügge, K. Werdan, H. W. Heldt. 1978. Specific transport of inorganic phosphate, 3-phosphoglycerate and triosephosphates across the inner membrane of the envelope in spinach chloroplasts. Biochim. Biophys. Acta 502: 232-247.

11a. Robinson, J. M., M. Gibbs, D. N. Cotler. 1977. Influence of pH upon the Warburg effect in isolated intact spinach chloroplasts. I. Carbon dioxide photoassimilation and glycolate synthesis. Plant Physiol. 59: 530-534.

12. Steup, M., D. G. Peavy, M. Gibbs. 1976. The regulation of starch metabolism by inorganic phosphate. Biochem. Biophys. Res. Commun. 72: 1554-1561.

13. Heldt, H. W., C. J. Chon, D. Maronde, A. Herold, Z.
    Stankovic, D. A. Walker, A. Kraminer, M. R. Kirk, U.
    Heber. 1977. Role of orthophosphate and other fac-
    tors in the regulation of starch formation in leaves
    and isolated chloroplasts. Plant Physiol.
    59: 1146-1155.

14. Walker, D. A. 1976. CO_2 fixation by intact chloro-
    plasts: Photosynthetic induction and its relation
    to transport phenomena and control mechanisms. Ref.
    10, pp. 235-278.

15. Cockburn, W., C. W. Baldry, D. A. Walker. 1967.
    Oxygen evolution by isolated chloroplasts with
    carbon dioxide as the hydrogen acceptor. A require-
    ment for orthophosphate or pyrophosphate. Biochim.
    Biophys. Acta 131: 594-596.

16. Sanwal, G. G., E. Greenberg, J. Hardie, E. C. Cameron,
    J. Preiss. 1968. Regulation of starch biosynthesis
    in plant leaves: Activation and inhibition of ADP-
    glucose pyrophosphorylase. Plant Physiol.
    43: 417-427.

17. Edwards, G. E., R. McC. Lilley, M. D. Hatch. 1979.
    Isolation of intact and functional chloroplasts from
    mesophyll and bundle sheath protoplasts of the C_4
    plant Panicum miliaceum. Plant Physiol. 63: 821-827.

18. Spalding, M. H., G. E. Edwards. 1980. Photosynthesis
    in isolated chloroplasts of the crassulacean acid
    metabolism plant Sedum praeltum. Plant Physiol.
    65: 1044-1049.

19. Robinson, S. P., J. T. Wiskich. 1977. Inhibition of
    CO_2 fixation by adenosine 5'-diphosphate and the role
    of phosphate transport in isolated pea chloroplasts.
    Arch. Biochem. Biophys. 184: 546-554.

20. Lilley, R. McC., A. W. D. Larkum. 1981. Isolation of
    functionally intact rhodoplasts from Griffithsia
    monilis (Ceramiaceae, Rhodophyta). Plant Physiol.
    67: 5-8.

21. Cockburn, W., C. W. Baldry, D. A. Walker. 1967.
    Photosynthetic induction phenomena in spinach chloro-
    plasts in relation to the nature of the isolating
    medium. Biochim. Biophys. Acta 143: 603-613.

22. Huber, S. C. 1979. Effect of photosynthetic inter-
    mediates on the magnesium inhibition of oxygen
    evolution by barley chloroplasts. Plant Physiol.
    63: 754-757.

23. Schwenn, J. D., R. McC. Lilley, D. A. Walker. 1973.
    Inorganic pyrophosphatase and photosynthesis by iso-
    lated chloroplasts. I. Characterization of chloro-
    plast pyrophosphatase and its relation to the
    response to exogenous pyrophosphate. Biochim.
    Biophys. Acta 325: 596-604.
24. Flügge, U. I., M. Freisl, H. W. Heldt. 1980. Balance
    between metabolite accumulation and transport in
    relation to photosynthesis by isolated spinach
    chloroplasts. Plant Physiol. 65: 574-577.
25. Robinson, S. P., J. T. Wiskich. 1977. p-Chloromer-
    curiphenyl sulphonic acid as a specific inhibitor of
    the phosphate translocator in isolated chloroplasts.
    FEBS Lett. 78: 203-206.
26. Walker, D. A., K. Kosciukiewicz, C. Case. 1973.
    Photosynthesis by isolated chloroplasts: Some fac-
    tors affecting induction in CO_2-dependent O_2 evolu-
    tion. New Phytol. 72: 237-247.
27. Huber, S. C. 1980. Effects of pH and other factors
    on the phosphate dependence of photosynthesis in
    spinach chloroplasts. Planta 149: 485-492.
28. Huber, S. C. 1979. Effect of pH on chloroplast
    photosynthesis. Inhibition of O_2 evolution by in-
    organic phosphate and magnesium. Biochim. Biophys.
    Acta 545: 131-140.
29. Lilley, R. McC., D. A. Walker. 1974. The reduction of
    3-phosphoglycerate by reconstituted chloroplasts and
    by chloroplast extracts. Biochim. Biophys. Acta.
    368: 269-278.
30. Stokes, D. M., D. A. Walker. 1972. Photosynthesis by
    isolated chloroplasts. Inhibition by DL-glyceralde-
    hyde of carbon dioxide assimilation. Biochem. J.
    128: 1147-1157.
30a. Kirk, M. R., U. Heber. 1976. Rates of synthesis and
    source of glycolate in intact chloroplasts. Planta
    132: 131-141.
31. Gimmler, H., G. Schäfer, H. Kraminer, U. Heber. 1974.
    Amino acid permeability of the chloroplast envelope
    as measured by light scattering, volumetry and
    amino acid uptake. Planta 120: 47-61.
32. Heldt, H. W., F. Sauer. 1971. The inner membrane of
    the chloroplast envelope as the site of specific
    metabolite transport. Biochim. Biophys. Acta
    234: 83-91.

33. Huber, S. C. 1979. Orthophosphate control of glucose-6-phosphate dehydrogenase light modulation in relation to the induction phase of chloroplast photosynthesis. Plant Physiol. 64: 846-851.

34. Walker, D. A. 1973. Photosynthetic induction phenomena and the light activation of ribulose diphosphate carboxylase. New Phytol. 72: 209-235.

35. Stitt, M., W. Wirtz, H. W. Heldt. 1980. Metabolite levels during induction in the chloroplast and extra chloroplast compartments of spinach protoplasts. Biochim. Biophys. Acta. 593: 85-102.

36. Buchanan, B. B. 1980. Role of light in the regulation of chloroplast enzymes. Annu. Rev. Plant Physiol. 31: 341-374.

37. Anderson, L. E. 1979. Interaction between photochemistry and activity of enzymes. Ref. 1, pp. 271-281.

38. Pradel, J., J. M. Soulie, J. Buc, J. C. Meunier, J. Ricard. 1981. On the activation of fructose-1,6-bisphosphatase of spinach chloroplasts and the regulation of the Calvin cycle. Eur. J. Biochem. 113: 507-511.

39. Anderson, L. E., S. C. Nehrlich, M. L. Champigny. 1978. Light modulation of enzyme activity: Activation of the light effect mediators by reduction and modulation of the enzyme activity by thiol-disulfide exchange. Plant Physiol. 61: 601-605.

40. Latzko, E., R. V. Garnier, M. Gibbs. 1970. Effect of photosynthesis, photosynthetic inhibitors and oxygen on the activity of ribulose 5-phosphate kinase. Biochem. Biophys. Res. Commun. 39: 1140-1144.

41. Anderson, L. E., M. Avron. 1976. Light modulation of enzyme activity in chloroplasts. Generation of membrane-bound vicinal dithiol groups by photosynthetic electron transport. Plant Physiol. 57: 209-213.

42. Wolosiuk, R. A, B. B. Buchanan. 1977. Thioredoxin and glutathione regulate photosynthesis in chloroplasts. Nature (London) 266: 565-567.

43. Wolosiuk, R. A., N. A. Crawford, B. C. Yee, B. B. Buchanan. 1979. Isolation of three thioredoxins from spinach leaves. J. Biol. Chem. 254: 1627-1632.

44. Lara, C., A. de la Torre, B. B. Buchanan. 1980. Ferralterin: An iron-sulfur protein functional in enzyme regulation in photosynthesis. Biochem. Biophys. Res. Commun. 94: 1337-1344.

45. Leegood, R. C., D. A. Walker. 1980. Autocatalysis and light activation of enzymes in relation to photosynthetic induction in wheat chloroplasts. Arch. Biochem. Biophys. 200: 575-582.

46. Charles, S. A., B. Halliwell. 1981. Light activation of fructose bisphatase in isolated spinach chloroplasts and deactivation by hydrogen peroxide. A physiological role for the thioredoxin system. Planta 151: 242-246.

47. Heldt, H. W., W. Laing, G. H. Lorimer, M. Stitt, W. Wirtz. 1981. On the regulation of CO_2 fixation by light. Proc. Fifth International Congress on Photosynthesis. In press.

48. Huber, S. C. 1978. Substrates and inorganic phosphate control the light activation of NADP-glyceraldehyde-3-phosphate dehydrogenase and phosphoribulokinase in barley (Hordeum vulgare) chloroplasts. FEBS Lett. 92: 12-16.

49. Furbank, R. T., R. McC. Lilley. 1980. Effects of inorganic phosphate on the photosynthetic carbon reduction cycle in extracts from the stroma of pea chloroplasts. Biochim. Biophys. Acta 592: 65-75.

50. Wolosiuk, R. A., M. E. Perelmuter, C. Chehebar. 1980. Enhancement of chloroplast fructose-1,6-bisphosphatase and dithiothreitol-reduced thioredoxin-f. FEBS Lett. 109: 289-293.

51. Woodrow, I. E., D. A. Walker. 1980. Light-mediated activation of stromal sedoheptulose bisphosphatase. Biochem. J. 191: 845-849.

52. Breazeale, V. D., B. B. Buchanan, R. A. Wolosiuk. 1978. Chloroplast sedoheptulose 1,7-bisphosphatase: Evidence for regulation by the ferredoxin/thioredoxin system. Z. Naturforsch. 33c: 521-528.

53. Levi, C., J. Preiss. 1978. Amylopectin degradation in pea chloroplast extracts. Plant Physiol. 61: 218-220.

54. Heldt, H. W., K. Werdan, M. Milovancev, G. Geller. 1973. Alkalization of the chloroplast stroma caused by light-dependent proton flux into the thylakoid space. Biochim. Biophys. Acta 314: 224-241.

55. Hall, D. O. 1976. The coupling of photophosphoryla-
    tion to electron transport in isolated chloroplasts.
    Ref. 10, pp. 135-170.

56. Werdan, K., H. W. Heldt, M. Milovancev. 1975. The
    role of pH in the regulation of carbon fixation in
    the chloroplast stroma. Studies on CO_2 fixation in
    the light and dark. Biochim. Biophys. Acta
    396: 276-292.

57. Purczeld, P., C. J. Chon, A. R. Portis, Jr., H. W.
    Heldt, U. Heber. 1978. The mechanism of the
    control of carbon fixation by the pH in the chloro-
    plast stroma. Studies with nitrite-mediated proton
    transfer across the envelope. Biochim. Biophys.
    Acta 501: 488-498.

58. Enser, U., U. Heber. 1980. Metabolic regulation by
    pH gradients. Inhibition of photosynthesis by
    indirect proton transfer across the chloroplast
    envelope. Biochim. Biophys. Acta. 592: 577-591.

59. Flügge, U. I., Freisl, H. W. Heldt. 1980. The
    mechanism of the control of carbon fixation by the
    pH in the chloroplast stroma. Studies with acid
    mediated proton transfer across the envelope.
    Planta 149: 48-51.

60. Zimmerman, G., G. J. Kelly, E. Latzko. 1976.
    Efficient purification and molecular properties of
    spinach chloroplast fructose 1,6-bisphosphatase.
    Eur. J. Biochem. 70: 361-367.

61. Avron, M., M. Gibbs. 1974. Carbon dioxide fixation
    in the light and dark by isolated spinach chloro-
    plasts. Plant Physiol. 53:140-143.

62. Huber, S. C. 1978. Regulation of chloroplast photo-
    synthetic activity by exogenous magnesium. Plant
    Physiol. 62: 321-325.

63. Demmig, B., H. Gimmler. 1979. Effect of divalent
    cations on cation fluxes across the chloroplast
    envelope and on photosynthesis of intact chloro-
    plasts. Z. Naturforsch. 34c: 233-241.

64. Gimmler, H., G. Schäfer, U. Heber. 1974. Low permea-
    bility of the chloroplast envelope towards cations.
    In Proc. Third International Congress on Photosyn-
    thesis (m. Avron, ed.). Elsevier, Amsterdam.
    pp. 1381-1392.

65. Portis, A. R., H. W. Heldt. 1976. Light-dependent changes of the Mg^{2+} concentration in the stroma in relation to the Mg^{2+} dependency of CO_2 fixation in intact chloroplasts. Biochim. Biophys. Acta 449: 434-446.

66. Huber, S. C., W. Maury. 1980. Effects of magnesium on intact chloroplasts. I. Evidence for activation of (sodium) potassium/proton exchange across the chloroplast envelope. Plant Physiol. 65: 350-354.

67. Maury, W. J., S. C. Huber, D. E. Moreland. 1981. Effects of magnesium on intact chloroplasts. II. Cation specificity and involvement of the envelope ATPase in (sodium) potassium/proton exchange across the envelope. Plant Physiol. In press.

68. Kaiser, W. M., W. Urbach, H. Gimmler. 1980. The role of monovalent cations for photosynthesis of isolated intact chloroplasts. Planta 149: 170-175.

69. Sokolove, P. M., T. V. Marsho. 1979. The effect of valinomycin on electron transport in intact chloroplasts. FEBS Lett. 100: 179-184.

70. Huber, S. C., W. J. Maury, D. E. Moreland. 1981. Further studies on the effects of Mg^2 on intact chloroplasts. Plant Physiol. 67 (Suppl.): 106.

71. Gimmler, H., B. Demmig, W. M. Kaiser. 1981. The role of K^+ and H^+-fluxes across the chloroplast envelope for photosynthetic CO_2-fixation. In Proc. Fifth International Congress on Photosynthesis (G. Akoynnoglou, ed). In press.

72. Enser, N., U. Heber. 1981. Maintenance of a pH gradient across the chloroplast envelope. Ref. 71, in press.

73. Kobayashi, Y., Y. Inoue, K. Shibata, U. Heber. 1979. Control of electron flow in intact chloroplasts by the intrathylakoid pH, not by the phosphorylation potential. Planta 146: 481-486.

74. Herold, A., D. H. Lewis. 1977. Mannose and green plants: Occurrence, physiology and metabolism, and use as a tool to study the role of orthophosphate. New Phytol. 79: 1-40.

75. Che-she, Sheu-Hwa, D. H. Lewis, D. A. Walker. 1975. Stimulation of photosynthetic starch formation by sequestration of cytoplasmic orthophosphate. New Phytol. 74: 383-392.

76. Giersch, C., U. Heber, G. Kaiser, D. A. Walker, S. P.
     Robinson. 1980. Intracellular metabolite gradients
     and flow of carbon during photosynthesis of leaf
     protoplasts. Arch. Biochem. Biphys. 205: 246-259.
77. Huber, S. C. 1981. Inter- and intra-specific
     variation in photosynthetic formation of starch
     and sucrose. Z. Pflanzenphysiol. 101: 49-54.
78. Huber, S. C. 1981. Interspecific variation in
     activity and regulation of leaf sucrose synthetase.
     Z. Pflanzenphysiol. 102: 443-450.

Chapter Six

CARBON METABOLISM IN GUARD CELLS

WILLIAM H. OUTLAW, JR.

Department of Biological Science (Unit 1)
Florida State University
Tallahassee, Florida 32306

INTRODUCTION

In the past decade stomatal physiology has been the subject of review articles,[1-17] two published symposia,[18,19] a multiauthored book,[20] and a forthcoming book,[21] all in addition to the older but excellent book of Meidner and Mansfield.[22] Two reviews[23,24] have been restricted to the topic of carbon metabolism alone. Thus, there are adequate general summaries, and it will not be my purpose to update these articles. Instead, I will presently "inventory" our current knowledge about carbon metabolism in guard cells. I will rely less on the authors' interpretations than on direct examination of the data and the methods used to obtain the data. With few exceptions (e.g. ref. 25), there were no quantitative studies on guard cell biochemistry prior to 1973. Since then, numerous reports have appeared. It is time to "take stock," so, to an extent, this is an internal report to stomatal biologists. As a result, I anticipate that readers will spend more time evaluating

the tabular data and less time with my narrative. I will
retrace the steps leading to our understanding the basic
outline of guard cell biochemistry involved in stomatal
movements. Therefore, to make the chronological sequence
clear, I have retained the author and the year of publication in the tabular data. This inventory will point to
weaknesses as well as strengths and, I hope, will suggest
some experiments and discourage others. In addition, I
have treated stomata of various species as if they were
identical. Certainly, as more work is reported, there will
be less and less justificaton for this assumption.

There are hazards in "taking stock"; it has forced me
to assess the confidence level of original reports. To
make this assessment, I have gone through the original
authors' methods and, where I thought appropriate, I have
recalculated all data on a common basis. To reduce the
potential impact of my mathematical errors or invalid
assumptions, I have included the original data and my
methods for conversion.

POSTULATED BIOCHEMISTRY

The biochemical steps postulated to occur in guard
cells during stomatal opening and closing are shown in
Figure 1. At least part of the cell turgor that causes
stomatal opening is a result of influx of potassium, which
results in an increase in concentration of several hundred
millimolar.[26-30] Presumably, cation influx is a result of
proton extrusion.[31] Details of this mechanism (e.g., elec-
trogenicity, energy coupling) need not concern us for the
present purposes. Importantly, inorganic anion uptake is
usually insufficient to balance cation influx[32-35] (but see
ref. 36). Ultimately, net proton efflux must be virtually
equal to the difference between K^+ and Cl^- uptake (see ref.
37 for mathematics). This difference is one-to-several-
hundred mmolar. Clearly, the cytoplasmic proton pool must
be replenished during stomatal opening to stabilize the pH
(presumably near 7.5). In broad terms, it is postulated
that starch is broken down to phosphoenolpyruvate. The
latter is carboxylated. During the biochemical steps, two
protons are released to the medium. Potassium accumulates
in the vacuole as an ionic solution of an organic anion salt
(e.g., K_2 malate) and as KCl. Alternatively, the organic
anion could be taken up from the surroundings instead of

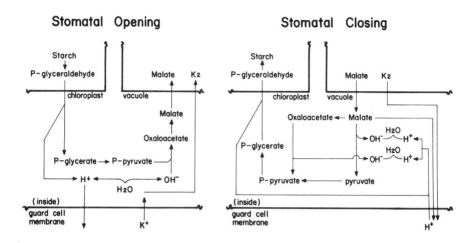

Figure 1. Biochemical steps postulated to occur in guard cells during stomatal opening and stomatal closing.

being synthesized internally. These hypotheses have been tested in a number of experiments, which will be described later.

During stomatal closure, if the organic anions (as well as K^+ and Cl^-) are lost from the guard cells, no mechanism is required to stabilize internal pH. However, organic anion excretion invokes compensatory carbon influx to replenish starch reserves for the next stomatal opening (if organic anions are synthesized internally during stomatal opening). Alternatively, part of the organic anions could be metabolized back to starch by guard cells themselves. These metabolic steps would require 2 protons/divalent anion. To a lesser extent than those for opening, these hypotheses have been tested. These experiments also will be described later.

For further details of the pH-stat mechanism, I refer the reader to the literature on cation uptake by roots (e.g., ref. 38) and more recent reviews.[39,40]

WHAT CONSTITUTES A PROPER EXPERIMENTAL SYSTEM TO STUDY GUARD CELL BIOCHEMISTRY?

Soon after the important role of potassium in stomatal movements was recognized, elegant methods (e.g., electron microprobe, ion-selective microelectrodes) were employed to localize and quantify potassium concentration in guard cells directly.[32,41-43] Unfortunately, this has not usually been the case for studies of carbon metabolism in guard cells. It is my feeling that much of the controversy in guard cell biochemistry has resulted from an improper choice of experimental system. To illustrate my point, I will give three examples of possible studies: (a) The first hypothetical study could be to determine the malate concentration in guard cells directly by using epidermal strips with open or closed stomata (Figure 2). If the strips are clean (i.e., free of adhering mesophyll cells), it should be possible to detect easily a six-fold change of malate concentration in guard cells if all epidermal cells are ruptured prior to the final analysis. However, if epidermal cells are not ruptured, an extrapolation of epidermal strip malate concentration to a guard cell basis probably should not be made. (b) The second hypothetical study could be directed toward determining enzyme activities (Figure 3) in guard cells. A preparation having 3% contamination by palisade cells (on a cell basis) yields an extract in which 50% of the chlorophyll is derived from the contaminant. Similar calculations could be done for epidermal strip preparations. In my opinion, the results of studies with epidermal strips that had not been demonstrated to be "completely" free of adhering mesophyll and to have had epidermal cells ruptured and extracted should not be extrapolated to guard cells (which are less than 5% of epidermis volume). (c) The third hypothetical study could be one of $^{14}CO_2$ metabolism by guard cell protoplasts. (Again, similar calculations could be made for epidermal strips.) Based on my calculations (Figure 4), a preparation with only 0.3% contamination (cell basis) would have equal amounts of ^{14}C incorporated by guard cells and by the palisade contamination. It seems clear that a valid interpretation of $^{14}CO_2$ incorporation by guard cells can be made only with stomatal systems demonstrated to be virtually free of any contamination.

Aside from the general comments made in this section, the appropriateness of the experimental system must be

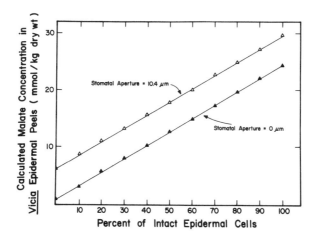

Figure 2. Apparent malate concentration in <u>Vicia</u> <u>faba</u> epidermal peel with open or closed stomata as a function of
epidermal cell intactness. For this graph, epidermal cells
were assumed to be 25 mmol/kg in malate.[44] This value was
assumed not to change with stomatal movements. Guard cells
of open and closed stomata were assumed to be 90 and 15
mmol/kg, respectively.[44] Guard cell pair dry mass was 6
ng.[44] Stomatal frequency ($52/mm^2$) was from Raschke and
Schnabl.[45] Dry mass/epidermal area ratio (4.43 µg/mm²) is
unpublished (Tarczynski, M. C., W. H. Outlaw, Jr.). Epidermal peels were assumed to have no adhering mesophyll cells.
Compare these values with those actually reported (Table 1).

judged by the individual reader. I ask the reader to bear
this section in mind when reviewing the tabular data.

EVIDENCE THAT GUARD CELLS ACCUMULATE MALATE DURING STOMATAL
OPENING

     Based on their direct determination of guard cell
potassium and chloride concentrations and their plasmolytic
studies, Humble and Raschke[32] suggested that (the equivalent of) an organic divalent anion would accumulate during
stomatal opening. Shortly thereafter Allaway[56] provided
evidence that this was the case by measuring malate in
"rolled" epidermal peels[43] of <u>Vicia</u> <u>faba</u> (Table 1) (In all
tables, I have selected one value for guard cell volume;

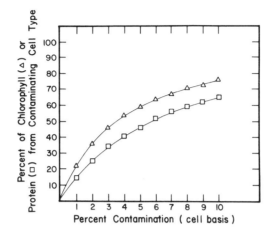

Figure 3. The relationship between contamination on a cell
basis and chlorophyll (Δ) or protein (□) basis. Although
this calculation was for Vicia faba guard cell protoplast
preparation contaminated with palisade cells, it is also
instructive in principal for guard cells in epidermal peels
where the tissue sample is a mixture of guard cells, epidermal cells, trichomes and adhering mesophyll cells.
Protein:chlorophyll ratios were from Outlaw et al.[46] (20:1
for palisade cells) and Outlaw et al.[47] (35:1 for guard
cells). Cell dry mass was from Outlaw and Lowry[44] (6
ng/guard cell pair) and Jones et al.[48] (12 ng/palisade
cell). Protein:dry mass ratios were from Outlaw et al.[49]
(5.2% and 21.1% of dry weight is protein in guard cells
and palisade cells, respectively).

this choice ignores changes in cell volume during stomatal
opening. However, the error is no greater than the
discrepancies among the volumes reported at various aper-
tures--see Fig. 4.3 of ref. 21). His results showed that
a pair of guard cells accumulate approximately 1 pmol of
malate during stomatal opening. This result has been confirmed by Ogawa et al.[60] and Raschke and Schnabl.[45] Our
measurements[44] of guard cell malate accumulation in situ
(using quantitative histochemical techniques[72-74]) showed
open stomata to contain about 0.5 pmol/guard cell pair. I
believe that all these are reliable results that can be
accepted at face value.

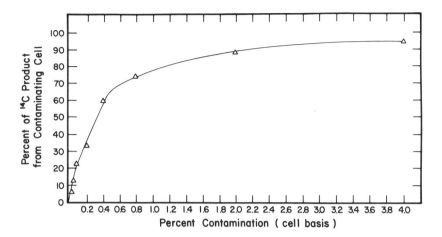

Figure 4. The relationship between contamination on a cell basis and the percent of $^{14}CO_2$ incorporated by the contaminating cells. These calculations were made for a guard cell protoplast preparation contaminated with palisade cells, but may provide insight for epidermal peels. See legend for Figure 3. To construct this hypothetical case, it was assumed that light stimulated CO_2 fixation in palisade cells by 200-fold (cf. refs. 50 and 51). Owing to the absence of the PCRP in <u>Vicia</u> guard cells,[52] it was assumed (for a first-order estimate) that light did not stimulate the rate of ^{14}C incorporation by guard cells. It was assumed that the rate of $^{14}CO_2$ fixation by guard cells was 9-fold higher (protein basis) than the dark rate for palisade cells. This last assumption was made because guard cells are enriched with PEPC.[53,54]

Schnabl[63] has reported that <u>Vicia</u> <u>faba</u> guard cell protoplasts accumulate up to 65 nmol malate/µg protein. Though I have calculated this value to be the equivalent of 20 pmol/guard cell pair, Schnabl (personal communication) has calculated it to be the equivalent of less than 1 pmol/guard cell pair. The difference in our calculations came from our different values for protein content. We[49] have determined a single <u>Vicia</u> <u>faba</u> guard cell to have approximately 0.16 ng protein while Schnabl (personal communication) found only 7 pg protein/guard cell protoplast.

Table 1. Malate concentration in epidermal tissue with open or closed stomata.

Reported Values		Difference	Percent Increase When Open	(Values Calculated by Reviewer; open/closed)		
Open Stomata (Aperture)	Closed Stomata (Aperture)			mmol/kg (dry wt)	mM	pmol/Guard Cell Pair
70 pmol/mm² ( )	0 pmol/mm² ( )	70 pmol/mm²	--	(187/0-guard cells) (15/0-whole epidermis)	200/0	1.1/0
9 µg/mg (10 µm)	3 µg/mg (0 µm)	6 µg/mg	200	(67/22)	--	--
11.5 µeq/0.1 gm (310 ppm CO₂)	6.5 µeq/0.1 gm (500 ppm CO₂)	5.0 µeq/0.1 gm	77	(58/33)	--	--
0.45 pmol/guard cell (12 µm)	0.04 pmol/guard cell (6 µm)	0.4 pmol/guard cell	1125	(150/15-guard cells) (5/0.5-whole epidermis)	--	(0.9/0.1)
80 mmol/kg (10.4 µm)	15 mmol/kg (0 µm)	65 mmol/kg	500	80/15	(48/10)	(0.5/0.1)
25 mmol/kg (10.4 µm)	18 mmol/kg (0 µm)	Not significant	--	22	--	--
30 nmol/mg (-)	10 nmol/mg (-)	20 nmol/mg	200	(200/65-guard cells) (30/10-whole epidermis)	(120/40)	(1.2/0.4)
250 pmol/mm² (12 µm)	37 pmol/mm² (2 µm)	213 pmol/mm²	576	(700/104-guard cells) (54/8-whole epidermis)	(420/60)	4.2/0.6
	13.4 pmol/mm² (4 m)	--	--	(42-guard cells) (3-whole epidermis)	(25)	0.25
147 peq/mm² (12 µm)	38 peq/mm² (6 µm)	109 peq/mm²	287	(233/60-guard cells) (16/4-whole epidermis)	(140/20)	1.4/0.2
65 nmol/µg protein (swollen)	20 nmol/µg protein (not swollen)	45 nmol/µg protein	225 (when swollen)	--	(3200/1940)	(20/6.2) (see text for discussion)
38 mmol/kg (-)	16 mmol/kg (-)	22 mmol/kg	138	38/16	--	--
3 µg/mg (25 µm)	4 µg/mg (0 µm)	-1 µg/mg	-25	(22/30)	--	--
5.5 µg/mg (12 µm)	2.7 µg/mg (0 µm)	2.8 µg/mg	104	(41/21)	--	--
64 mol/m³ (-)	61 mol/m³ (-)	3 mol/m³	5	--	64/61	--
40 nmol/cm² (-)	10 nmol/cm² (-)	30 nmol/cm²	300	(140/35)	--	--
185 pmol/mm² (16 µm)	15 pmol/mm² (2 µm)	170 pmol/cm²	1119	(64/5)	(240/50)	(3.4/0.3)
	29.7 pmol/mm² (4 µm)	--	--	--	--	0.25

Species	Tissue System	Comments	Reference
Vicia faba	"rolled" epidermal peels[43]	Dry mass/area ratio (0.46 mg/cm^2);[57] guard cell pair mass (6 ng).[44]	Allaway, 1973[56]
Vicia faba	epidermal peels	Values are from his Fig. 1 & 2.	Pearson, 1973[57]
Vicia faba	epidermal peels		Pallas & Wright, 1973[58]
Vicia faba	"rolled" epidermal peels[43]	See 1st entry under Allaway, 1973;[56] stomatal frequency (52/mm^2).[45]	Allaway, 1976[1]
Vicia faba	guard cells	Values are from their Fig. 2. Guard cell pair mass (6 ng) from their Table 1; guard cell volume (5 pl).[59]	Outlaw & Lowry, 1977[44]
Vicia faba	epidermal cells		Outlaw & Lowry 1977[44]
Vicia faba	sonicated epidermal peels	Values shown are from their Fig. 1 after 4 hr of light or darkness. Conversion to area basis from their text (0.2 mg/cm^2). Stomatal frequency (52/mm^2);[45] guard cell volume (5 pl);[59] guard cell pair dry mass (6 ng).[44]	Ogawa et al., 1978[60]
Vicia faba	epidermal peels	See entry under Schnabl, 1978.[62] Conversion to stoma basis by original authors because most epidermal cells were ruptured during peeling (60 stomata/mm^2). Guard cell volume (5 pl).[59] Strips were floated on KIDA.	Van Kirk & Raschke, 1978[61]
Vicia faba	epidermal peels	Dry mass/area ratio (0.46 mg/cm^2);[57] values given were those from plants grown on standard Cl$^-$ nutrient (her table 2); conversion to stoma basis by original author because most epidermal cells were stated to be ruptured during peeling; guard cell volume (5 pl);[59] dry mass/guard cell (3 ng).[44]	Schnabl, 1978[62]
Vicia faba	epidermal peels	Dry mass/area ratio (0.46 mg/cm^2);[57] values given were those reported for -Cl$^-$ medium; conversion to stoma basis possible because authors state that most epidermal cells were ruptured during incubation.	Raschke & Schnabl, 1978[45]
Vicia faba	guard cell protoplasts	Protein content (5.2%);[49] guard cell mass (3 ng);[44] protoplast volume (3.2 & 1.6 pl) from her text.	Schnabl, 1980[63]
Vicia faba	epidermal peels	Unpublished data (Brooks, J. M., W. H. Outlaw, Jr.).	
Commelina cyanea	epidermal peels	Values from his Fig. 1 & 2.	Pearson, 1973[57]
Commelina cyanea	epidermal peels	Dry mass/area ratio (0.29 mg/cm^2);[57] Values used here are estimated from their figure 7.	Pearson & Milthorpe, 1974[64]
Commelina communis	epidermal peels		Bowling, 1976[65]
Commelina communis	"clean" epidermal peels	Dry mass/area ratio of C. cyanea (0.29 mg/cm^2).[57]	Raschke & Dittrich, 1977[67]
Commelina communis	"isolated" guard cells[69]	Dry mass/area ratio of C. cyanea (0.29 mg/cm^2);[57] Stomatal frequency (5386/cm^2);[68] guard cell volumes for C. cyanea (7.2 & 3.2 pl).[64]	Travis and Mansfield, 1977[66]
Allium cepa	epidermal peels		Schnabl, 1978[62]

Table 1. (Continued)

Reported Values			Percent	(Values Calculated by Reviewer; open/closed)		
Open Stomata (Aperture)	Closed Stomata (Aperture)	Difference	Increase When Open	mmol/kg (dry wt)	mM	pmol/Guard Cell Pair
4 nmol/µg protein (swollen)	5 nmol/µg prot protein (not swollen)	Not significant	0	--	--	--
3.44 mg/gm		Not significant	0	(26)	--	--
10.5 µeq/gm fresh (-)	10.6 µeq/gm fresh (-)	-0.1 µeq/gm fresh	-1	(35/35)	--	--

Independent confirmation of one of these values will be necessary for further interpretation.

A number of studies in which the epidermal peel malate concentration did not change with stomatal aperture have been reported (Table 1). There are several possible explanations (e.g., involvement of Cl^- as a counter anion to K^+,[61] differences among species,[71] malate transport from epidermal to guard cells,[65] etc.). However, until it is shown that malate does not accumulate in guard cells when a proper experimental system is used, I think that the conclusion that guard cells accumulate malate upon stomatal opening should be regarded as valid.

EVIDENCE THAT GUARD CELLS ACCUMULATE ORGANIC ANIONS OTHER THAN MALATE DURING STOMATAL OPENING

Direct measurements[44] of citrate in guard cells of Vicia faba showed an increase of 0.35 pmol/guard cell pair when stomata were open (Table 2). Thus, accumulation of citrate also provides charge balancing equivalents. Organic anions other than malate and citrate have not been demonstrated to contribute significantly to balancing K^+ influx. However, significant elevations in guard cell glutamate and aspartate concentrations of open stomata do potentially point to higher rates of acid metabolism.

Species	Tissue System	Comments	Reference
Allium cepa	guard cell protoplasts		Schnabl, 1980[63]
Tulipa gesneriana	epidermal peels	Peels stated to be free of underlying mesophyll tissues. Data given are pooled data from their Table 1. They state that no significant differences in malate were found when stomata were open or closed.	Rutter et al., 1977[70]
Pelargonium X hortorum	epidermal peels	Dry weight was arbitrarily calculated as 15% of fresh weight.	Contour-Ansel & Longuet, 1979[71]

EVIDENCE THAT ORGANIC ANIONS ARE SYNTHESIZED BY GUARD CELLS DURING STOMATAL OPENING

There are several lines of evidence that malate is synthesized in guard cells and is not imported. A direct line of evidence[55] is that malate is accumulated in guard cells in "rolled" epidermal strips.[43] The epidermal cells of these peels had been ruptured and the contents washed away before stomatal opening was induced by light. These results conclusively show that import of malate is not required for guard cell malate accumulation during stomatal opening in Vicia faba. In addition, there are three indirect lines of evidence that malate is synthesized in guard cells: (a) Guard cells have high levels of phospho-enolpyruvate carboxylase (PEPC) activity (Table 3). (b) The decrease in starch concentration (Table 4) was exactly that required for anion synthesis.[81] (However, the glucose released from starch breakdown could have been used for sucrose synthesis.) (c) $^{14}CO_2$ fixation by stomatal systems results in ^{14}C-malate synthesis (Table 5). However, the absolute increase in chemical pool size was found to be about 300x too low[67] to account for malate accumulation in the one case I am aware of in which the investigators have measured malate accumulation and ^{14}C incorporation. Other laboratories should repeat these measurements.

To my knowledge, there is no experimental evidence to support Bowling's hypothesis[65] that malate is imported into guard cells during stomatal opening.

Table 2.   Concentrations of organic anions (excluding malate) in epidermal tissue with open or closed stomata.

Anion	Reported Values		Difference	Percent Increase When Open
	Open Stomata (Aperture)	Closed Stomata (Aperture)		
aspartate	12 mmol/kg (10.4 µm)	7 mmol/kg (0 µm)	5 mmol/kg	71
	15 mmol/kg (10.4 µm)	12 mmol/kg (0 µm)	Not significant	0
citrate	42 µeq/0.1 gm (310 ppm CO_2)	24 µeq/0.1 gm (500 ppm CO_2)	18 µeq/0.1 gm	75
	100 mmol/kg (10.4 µm)	40 mmol/kg (0 µm)	60 mmol/kg	150
	260 mmol/kg (10.4 µm)	165 mmol/kg (0 µm)	95 mmol/kg	58
glutamate	13 mmol/kg (10.4 µm)	6 mmol/kg (0 µm)	7 mmol/kg	117
	18 mmol/kg (10.4 µm)	18 mmol/kg (0 µm)	0 mmol/kg	0
glycerate	38 µeq/0.1 gm (310 ppm CO_2)	24 µeq/0.1 gm (500 ppm CO_2)	16 µeq/0.1 gm	67
	10 mmol/kg (10.4 µm)	17 mmol/kg (0 µm)	Not significant	0
	23 mmol/kg (10.4 µm)	44 mmol/kg (0 µm)	-21mmol/kg	-48
isocitrate	4 mmol/kg (10.4 µm)	3 mmol/kg (0 µm)	Not significant	0
	4 mmol/kg (10.4 µm)	4 mmol/kg (0 µm)	0	0
succinate	0.05 mg/gm		0	0
fumarate	1.93 mg/gm		0	0

EVIDENCE THAT THE PHOTOSYNTHETIC CARBON REDUCTION PATHWAY (PCRP) IS ABSENT IN GUARD CELLS

Whether the PCRP is operative in guard cells is an important question. If present, the PCRP could be a source of reduced carbon during stomatal opening in light.   In addition, the PCRP potentially could be part of a CO_2 sensing mechanism (because of its relationship to glycolate metabolism[88,89]).   However, if present, the PCRP would cause the elevation of the P-glycerate concentration in light.[90]   Elevated P-glycerate concentration is a positive signal for starch synthesis (through activation of ADP-glucosepyrophosphorylase[91]).   Normally, guard cells

(Values Calculated by Reviewer; open/closed)

mmol/kg (dry wt)	mM	pmol/Guard Cell Pair	Species	Tissue System	Comments	Reference
12/7	.(7/4)	(0.07/0.04)	Vicia faba	guard cells	Values from their Fig. 2. Guard cell pair dry mass (6 ng) from their Table 1. Guard cell volume (5 pl).[59]	Outlaw & Lowry, 1977[44]
15/12	--	--	Vicia faba	epidermal cells		Outlaw & Lowry, 1977[44]
(140/80)	--	--	Vicia faba	epidermal peels		Pallas & Wright 1973[58]
100/40	(60/24)	(0.6/0.24)	Vicia faba	guard cells	See 1st entry under Outlaw and Lowry, 1977.[44]	Outlaw & Lowry, 1977[44]
260/165	--	--	Vicia faba	epidermal cells		Outlaw & Lowry, 1977[44]
13/6	(8/4)	(0.08/0.04)	Vicia faba	guard cells	See 1st entry under Outlaw and Lowry, 1977.[44]	Outlaw & Lowry, 1977[44]
18/18	--	--	Vicia faba	epidermal cells		Outlaw & Lowry, 1977[44]
(380/240)	--	--	Vicia faba	epidermal peels		Pallas & Wright, 1973[58]
10/17	(6/10)	(0.06/0.1)	Vicia faba	guard cells	See 1st entry under Outlaw and Lowry, 1977.[44]	Outlaw & Lowry, 1977[44]
23/44	--	--	Vicia faba	epidermal cells		Outlaw & Lowry, 1977[44]
4/3	(2/2)	0.02/0.02)	Vicia faba	guard cells	See 1st entry under Outlaw and Lowry, 1977.[44]	Outlaw & Lowry, 1977[44]
4/4	--	--	Vicia faba	epidermal cells		Outlaw & Lowry, 1977[44]
(0.4)	--	--	Tulipa gesneriana	epidermal peels	Pooled data from their table 1. They found no differences in the levels of organic acids from epidermis with open and closed stomata.	Rutter et al., 1977[70]
(16)	--	--	Tulipa gesneriana	epidermal peels	See entry under Rutter et al., 1977.[70]	Rutter et al., 1977[70]

degrade starch as stomata are opened by light (Table 4).
Thus the presence of the PCRP in guard cell chloroplasts
would appear to present a paradox.

The question of the presence or absence of the PCRP in
guard cells has been one of the most controversial aspects
of stomatal physiology. Investigators have approached this
problem in two ways: (a) by determination of the products
of $^{14}CO_2$ fixation by stomatal systems (Table 5) and (b) by
activity measurements of enzymes of the PCRP (Table 6).
Both of these methods are susceptible to artifactual results
owing to the relatively enormous impact of contamination
(Fig. 3, 4). The effect of contamination in $^{14}CO_2$ fixation
studies (Fig. 4) is especially difficult to avoid.

Table 3. Reported activities of phosphoenolpyruvate carboxylase in epidermal tissue.

Reported activity	% of "Mesophyll" on Basis Reported	PEPC/ Rubisco	(Values Calculated by Reviewer)		
			μmol/mg chl·hr	μmol/mg protein·hr	pmol/Guard Cell Pair·hr
1 pmol/guard cell·hr	mesophyll not given	--	(222)	(6.6)	2
650 mmol/kg·hr	217	>130	(438)	(12.5)	(3.9)
2200 μmol/mg chl·hr	31500	629	2200	(63)	(20)
280 mmol/kg·hr	93	>7	--	(3.8)	--
5.2 μmol/mg protein·hr	91	26	(1898)	5.2	--
751 μmol/mg chl·hr	2276 (Chl); 57 (protein)	6.5 (chl); 8.7 (prot)	751	7.0	--
8.7/0.86 μmol/mg protein·hr	1913 (Chl); 65-125 (protein)	8-80	(1392/138)	8.7/0.86	--
38960 cpm/3 cm^2·10 min	mesophyll not given	18	(ca. 13)	(ca. 0.1)	(ca. 0.3)
1125 μmol/mg chl·hr	3041	--	1125	(7.8)	--
11200 μmol/mg chl·hr	10769 (chl); 95 (protein)	8.4 (chl); 6.6 (protein)	11200	9.4	--
6253 μmol/mg chl·hr	6381	--	6381	--	--

All studies prior to 1977 indicated that the PCRP was present in guard cells (Table 5, 6). Since then, two laboratories[67,84,92] have reported the inability to detect radioactive PCRP intermediates in extracts of stomatal systems that had been incubated with $^{14}CO_2$. Raschke and Dittrich[67] (Table 5) reported only 3 soluble compounds (malate, aspartate and citrate) to have incorporated ^{14}C when "clean" Commelina communis epidermal strips were incubated with $^{14}CO_2$ for 0.5 to 6 min in light (their table 5). The rate of ^{14}C incorporation (50 cpm/mm^2·min, their table 3)

Species	Tissue System	Comments	Reference
Vicia faba	"rolled" epidermal peels[43]	Chl:guard cell pair (9 pg);[47,75] protein content (5.2%).[49]	Allaway, 1976[1]
Vicia faba	guard cells	Protein:Chl ratio (35:1);[47] Protein content (5.2%);[49] Guard cell pair dry mass (6 ng).[44]	Outlaw & Kennedy, 1978[53]
Vicia faba	guard cell protoplasts	See 1st entry under Outlaw Kennedy, 1978.[53]	Schnabl, 1981[80]
Vicia faba	epidermal cells	Protein content (7.1%).[49]	Outlaw & Kennedy, 1978[53]
Commelina cyanea	epidermal peels	Protein:Chl ratio (365:1) derived from their text.	Thorpe et al., 1978[76]
Commelina communis	epidermal peels		Willmer et al., 1973[77]
Commelina communis	epidermal peels	1st/2nd value is V_{max} from peel illuminated/darkened before extraction; protein: chl ratio (160:1) calculated from their report. These values are from their Table 1 and Figures 2 & 3.	Donkin & Martin, 1980[78]
Commelina communis	"clean" epidermal peels	Calculated figures are "ballpark" owing to the circuitous route of calculation: cpm for malate + aspartate synthesized given; specific activity of $^{14}CO_2$ given; counting efficiency (ca. 40%) calculated internally from their report; area/mass ratio for C. cyanea (0.29 mg/cm^2;)[57] chl:mass for C. cyanea (0.41 mg chl/gm);[76] protein:chl (107:1);[77] stomatal frequency (5386/cm^2).[68]	Raschke & Dittrich, 1977[67]
Commelina benghalensis	epidermal peels	Values in parentheses calculated by Thorpe et al., 1978[76]	Rama Das & Raghavendra, 1974[79]
Tulipa gesneriana	epidermal peels		Willmer et al., 1973[77]
Tridax procumbens	epidermal peels		Rama Das & Raghavendra, 1974[79]

was low, which resulted in modest total ^{14}C in their samples (25 to 100 cpm for their 0.5-min data point; 1 to 4 cm^2 were used per treatment (their methods)). Thus, ^{14}C-intermediates of the PCRP would have been difficult to detect among the chromatographed products. Nevertheless, their report[67] was an extremely valuable one because it was the first indication, to my knowledge, that chloroplast-containing guard cells lack the PCRP. Schnabl[84,92] incubated guard cell protoplasts with $^{14}CO_2$; she also did not detect ^{14}C-intermediates of the PCRP. However, 24% of the ^{14}C

Table 4.   Carbohydrate concentration in epidermal tissue with open or closed stomata.

Compound	Reported Values Open Stomata (Aperture)	Closed Stomata (Aperture)	Difference	Percent Increase When Open
Starch	87 mmol/kg (ca. 10.4 µm)	159 mmol/kg (0 µm)	-72 mmol/kg	-45
	10 mmol/kg (ca. 10.4 µm)	10 mmol/kg (0 µm)	0	0
	16% light abs (light, -CO$_2$)	24% light abs (dark, +CO$_2$)	-8% light abs	-50
Sucrose	0.7 mg/0.1 gm (160 ppm CO$_2$)	0.8 mg/0.1 gm (310 ppm CO$_2$)	Not Significant	0
	76 µg/mg (11 µm)	30 µg/mg (0 µm)	46 µg/mg	153
	117 mmol/kg (ca. 10.4 µm)	72 mmol/kg (0 µm)	45 mmol/kg	62
	91 mmol/kg (ca. 10.4 µm)	46 mmol/kg (0 µm)	45 mmol/kg	100
	2 µg/mg (25 µm)	0 µg/mg (0 µm)	2 µg/mg	--
	13.1 mg/gm		0	0
fructose	2.42 mg/gm		0	0
glucose	1.0 mg/0.1 gm (160 ppm CO$_2$)	0.8 mg/0.1 gm (310 ppm CO$_2$)	Not Significant	0
	25.1 mg/gm		0	0

incorporated was in glycolate (an intermediate of the Photorespiratory Carbon Oxidation Pathway (PCOP)).   The finding of such high levels of ^{14}C in glycolate is difficult to reconcile with our current knowledge[93] about the relationship between the PCRP and the PCOP.   This potential difficulty is compounded by our failure (R. Hampp et al., unpublished) to detect hydroxypyruvate reductase in Vicia guard cells.   However, these interpretations must be qualified. ^{14}C-glycolate may be synthesized by alternate pathways (e.g., through conversion of isocitrate to glyoxylate,[94] which can be reduced to glycolate,[95] or by biochemical steps beginning with the hydrolysis of P-glycerate[96]).   Three

| (Values Calculated by Reviewer; open/closed) | | | | | | |
mmol/kg (dry wt)	mM	pmol/Guard Cell Pair	Species	Tissue System	Comments	Reference
87/159	--	(0.5/0.9)	Vicia faba	guard cells	Starch concentration in anhydroglucosyl equivalents. Guard cell pair mass (6 ng).[44]	Outlaw & Manchester, 1979[81]
10/10	--	--	Vicia faba	epidermal cells		Outlaw & Manchester, 1979[81]
--	--	--	--	guard cells	Average of several species.	Mouravieff, 1972[82]
(22)	--	--	Vicia faba	epidermal peels	Typical data (II-71, 72) from their Table I.	Pallas & Wright, 1973[58]
(223/88)	--	--	Vicia faba	epidermal peels	Values from his Fig. 1 & 2.	Pearson, 1973[57]
117/72	(70/43)	(0.7/0.4)	Vicia faba	guard cells	See 1st entry under Outlaw and Manchester, 1979.[81] Guard cell volume (5 pl).[59]	Outlaw & Manchester, 1979[81]
91/46	--	--	Vicia faba	epidermal cells		Outlaw & Manchester, 1979[81]
(6/0)	--	--	Commelina cyanea	epidermal peels		Pearson, 1973[57]
(38)	--	--	Tulipa gesneriana	epidermal peels	Pooled data from their Table 2. They found no difference in level of sugars in epidermis with open or closed stomata.	Rutter et al., 1977[70]
(13)	--	--	Tulipa gesneriana	epidermal peels	See 1st entry under Rutter et al., 1977.[70]	Rutter et al., 1977[70]
(50)	--	--	Vicia faba	epidermal peels	See 1st entry under Pallas & Wright, 1973.[58]	Pallas & Wright, 1973[58]
(139)	--	--	Tulipa gesneriana	epidermal peels	See 1st entry under Rutter et al., 1977.[70]	Rutter et al., 1977[70]

other laboratories have determined the products of $^{14}CO_2$ fixation in "clean" guard cell preparations. Willmer et al.[87] and Thorpe et al.[97] screened their epidermal peels for mesophyll contamination before labelling. Allaway[1] "rolled" his epidermal strips,[43] but subsequent work[55] with autoradiography has revealed the strips to have mesophyll contamination. We (P. H. Brown et al., unpublished) examined our Vicia guard cell protoplasts preparations before labelling, and estimate contamination (cell basis) to be about 0.1% or less. ^{14}C-intermediates of the PCRP were detected in all these last three studies. There are two possible explanations for these positive results, which

Table 5. Distribution of ^{14}C among products during steady-state labeling of plant tissue with $^{14}CO_2$ ($H^{14}CO_3^-$) in light.

	8 sec	15 sec				20 sec	30 sec						60 sec													
	A	B_1	K_1	L_1	H_1	M_1	B_2	E_1	I_1	J_1	K_2	L_2	B_3	C	D	E_2	F	G	H_2	I_2	J_2	K_3	L_3	M_2	N	O
ala	1	-	0	0	-	-	-	1	0	-	0	0	-	-	-	1	0	-	-	-	-	1	1	-	-	-
gly+ser+gln	3	-	4	19	0	-	-	1	-	-	2	3	-	25	9	1	1	-	-	-	-	1	2	-	<3	-
asp	-	-	19	22	0	-	-	36	24	38	19	27	-	-	14	34	17	-	-	19	30	34	32	-	36	-
asn	-	-	-	0	-	-	-	<0.5	-	-	-	-	-	-	-	<0.5	-	-	-	-	-	-	-	-	-	-
glu	-	-	-	-	-	-	-	1	0	-	1	2	-	-	-	1	0	-	-	-	-	0	3	-	-	-
total basic	3	8	23	41	16	12	10	39	24	38	22	32	17	25	23	37	17	-	28	19	30	36	38	22	39	-
sucrose	1	-	0	0	-	-	-	1	-	-	0	0	-	-	-	5	-	-	-	-	-	1	0	-	-	-
maltose	-	-	-	-	-	-	-	<0.5	-	-	-	-	-	-	-	<0.5	-	-	-	-	-	-	-	-	-	-
glucose	-	-	-	-	-	-	-	<0.5	-	-	-	-	-	-	-	<0.5	-	-	-	-	-	-	-	-	-	-
fructose	-	-	-	-	-	-	-	<0.5	-	-	-	-	-	-	-	<0.5	-	-	-	-	-	-	-	-	-	-
total neutral	1	3	0	0	11	0	5	1	0	-	0	0	9	0	0	5	0	-	5	0	-	1	0	0	5	-
malate	-	-	65	51	20	81	-	29	76	-	69	55	-	30	26	24	26	52	59	81	-	56	50	68	48	80
citrate	-	-	0	3	-	-	-	2	-	-	2	3	-	12	4	3	0	-	-	-	-	3	3	-	-	-
fumerate	-	-	-	-	-	-	-	4	-	-	-	-	-	-	-	1	-	-	-	-	-	-	-	-	-	-
pyruvate	-	-	-	-	-	-	-	<0.5	-	-	-	-	-	-	-	<0.5	-	-	-	-	-	-	-	-	-	20
glycollate	2.5	-	-	-	-	-	-	<0.5	-	-	-	-	-	24	0	-	22	-	-	-	-	-	1	-	-	-
total organic acid	2.5	26	65	54	20	81	16	35	76	-	72	61	9	66	30	28	48	52	59	81	-	60	54	68	48	100
PGA	<0.7	-	0	0	-	-	-	6	0	-	0.5	1	-	5	14	4	-	-	-	-	-	<0.5	1	2	-	-
PEP	-	-	0	0	-	-	-	<0.5	<0.5	-	-	-	-	-	<0.5	<0.5	-	-	-	-	-	1	-	-	-	-
Sugar-P	90	-	0	0	-	-	-	10	0	-	0	1	-	4	33	23	-	-	-	0	-	1	4	2	8	-
total phosphorylated	90	38	0	0	41	0	41	16	0	-	0.5	2	26	9	47	27	-	-	8	0	-	1	5	4	8	-
other (insol or starch)	-	25	6	3	-	-	28	3	-	-	1	1	39	0	0	3	35	48	-	-	-	1	1	-	-	-
total	97	100	94	98	88	93	100	94	100	38	96	96	100	100	100	100	100	100	100	100	30	99	98	94	100	100
100% =	H_2O-sol	all	all	all	80% EtOH sol	75% EtOH sol	all	70% EtOH sol	80% MeOH sol	80% MeOH sol	all	all	all	H_2O-sol	H_2O-sol	all	H_2O-sol	H_2O-sol	80% EtOH sol	70% EtOH sol	80% MeOH sol	all	all	75% EtOH sol	80% MeOH sol	75% MeOH sol

(In the "other" row, the F and G values of 35 and 48 are labeled "mucilage".)

A – Palisade cells of Vicia faba, in situ (Outlaw & Fisher, 1975);[83]
B – Palisade cell protoplasts of Vicia faba (Outlaw et al., 1976);[46]
C – Mesophyll cell protoplasts of Vicia faba (Schnabl, 1980);[84]
D – Mesophyll cell protoplasts of Allium cepa (Schnabl, 1980);[84]
E – Guard cell protoplasts of Vicia faba (P. H. Brown et al., unpublished)
F – Guard cell protoplasts of Vicia faba (Schnabl, 1980);[84]
G – Guard cell protoplasts of Allium cepa (Schnabl, 1980);[84]
H – "Rolled" epidermal peel of Vicia faba (Allaway, 1976);[1]
I – "Clean" epidermal peel of Commelina communis (Raschke & Dittrich,1977);[67]

J – "Clean" epidermal peel of Commelina cyanea (Thorpe et al., 1978);[76]
K – Epidermal peel (with open stomata) of Tulipa gesneriana (Raschke & Dittrich, 1977);[67]
L – Epidermal peel (with closed stomata) of Tulipa gesneriana (Raschke & Dittrich, 1977);[67]
M – Epidermal peel of Commelina diffusa (Willmer & Dittrich, 1974);[86]
N – Clean epidermal peel of Commelina cyanea (Willmer et al., 1978);[87]
O – Epidermal peel of Allium cepa (Schnabl, 1977);[85]

are consistent with the absence of the PCRP in guard cells:
(a) sugar-P would be expected to be labeled with ^{14}C through
the gluconeogenic pathway anyway and (b) extremely low
levels of contamination by mesophyll chloroplasts, which are
difficult to detect with transmitted-light microscopy,[98]
might have been responsible (see Fig. 4).

Future reports of $^{14}CO_2$ fixation studies with stomatal
systems should include estimates of contamination, should
directly report the total ^{14}C contents of the sample, and
should be of shorter duration experiments (1 to 3 sec).

Several laboratories have reported high levels of
ribulose-1,5-bisphosphate carboxylase (Rubisco) in extracts
of stomatal systems (Table 6). However, only one[76] of the
positive reports verified that contamination was insignificant. That positive report was by Thorpe et al.,[76] who
reported a PEPC/Rubisco ratio of 26:1 on a protein basis
(Table 3). With "clean" epidermal peels of C. communis,
Raschke and Dittrich[67] found low levels of Rubisco (PEPC/-
Rubisco = 18:1, extract basis (Table 3)). We[52] were unable
to detect Rubisco activity (PEPC/Rubisco > 130, Table 3) in
Vicia faba guard cells by histochemical techniques or in
extracts of guard cell protoplasts. The last has recently
been confirmed.[80] In addition, immunoelectrophoresis
against Rubisco in extracts of guard cell protoplasts was
negative.[52] Other PCRP enzymes were also not detected in
guard cells of Vicia faba[52] (Table 6). We (unpublished)
have failed to detect Rubisco in guard cells of several
other species as well. In work yet to be published,
Madhavan and Smith[99] attempted to detect Rubisco in guard
cells of several species using an indirect, immunofluorescence technique. Except for guard cells of a CAM
plant, their results were negative. Independent confirmation of the positive result will be important as it will
bear directly on the regulation of Rubisco gene expression.
It is tempting to speculate that Rubisco genes are expressed in guard cells of CAM plants (in which the stomata open
nocturnally) because its presence does not conflict with
starch degradation upon stomatal opening.

In summary, most of the recent reports, which were with
"pure" stomatal systems, indicate--to the degree that they
are reliable--the absence of the PCRP in guard cell chloro-

Table 6. Reported activities of enzymes of the photosynthetic carbon reduction pathway in epidermal tissue.

Enzyme	Reported Activity	% of Mesophyll on Basis Reported	(Values Calculated by Reviewer)		
			μmol/mg chl·hr	μmol/mg protein·hr	pmol/Guard Cell Pair·hr
Rubisco	N.D. (<5mmol/kg·hr)	< 0.4	(< 4)	(< 0.1)	(< 0.03)
	N.D. (<4 μmol/mg chl·hr)	< 1	< 4	(< 0.1)	(< 0.04)
	3.5 μmol/mg chl·hr	< 0.2	3.5	(0.1)	(0.03)
	< 40 mmol/kg·hr	< 3	--	(< 0.5)	--
	0.2 μmol/mg protein·hr	1	(73)	0.2	--
	115 μmol/mg chl·hr	74 (chl); 1 (protein)	115	0.8	--
	18.2 μmol/mg chl·hr	44 (chl); 10 (protein)	18.2	0.1	--
	2153 cpm/3 cm^2·10 min	Mesophyll not given	(ca. 0.7)	(ca. 0.06)	(ca. 0.02)
	1340 μmol/mg chl·hr	123 (chl); 1 (protein)	1340	1.4	--
Ru-5-P Kinase	N.D. (<260 mmol/kg·hr)	< 1.5	(< 175)	(< 5)	(< 1.5)
	N.D. (<123 mmol/kg·hr)	< 0.7	--	(< 2.4)	--
(NADP$^+$) Glyceraldehyde-P	N.D. (<27 mmol/kg·hr)	< 2	(< 18)	(< 0.5)	(< 0.2)
Dehydrogenase	N.D. (< 31 mmol/kg·hr)	< 3	--	(< 0.4)	--

plasts. However, at present, it is premature to generalize
these results to all species.

EVIDENCE FOR CHANGES IN CARBOHYDRATE CONCENTRATION IN GUARD CELLS DURING STOMATAL OPENING

The concentration of sucrose has been reported to be
higher in open stomatal systems than in closed ones[25]
(Table 4). However, there has not been general agreement
on this point (Table 4). In those reports that showed an

Species	Tissue System	Comments	Reference
Vicia faba	guard cells	Prot:chl ratio (35:1);[47] protein content (5.2%).[49]	Outlaw et al., 1979[52]
Vicia faba	guard cell protoplasts	See 1st entry under Outlaw et al., 1979;[52] guard cell pair mass (6 ng).[44]	Outlaw et al., 1979[52]
Vicia faba	guard cell protoplasts	See 1st entry under Outlaw et al., 1979.[52]	Schnabl, 1981[80]
Vicia faba	epidermal cells	Protein content (7.1%).[49]	Outlaw et al., 1979[52]
Commelina cyanea	epidermal peels	Protein:chl ratio (365:1) derived from their text.	Thorpe et al., 1978[76]
Commelina communis	epidermal peels		Willmer et al., 1973[77]
Commelina communis	epidermal peels	These values are from their table 1.	Donkin & Martin, 1980[78]
Commelina communis	"clean" epidermal peels	The calculated figures are "ballpark" owing to the circuitous route of calculation: cpm for PGA synthesized given; specific activity of $^{14}CO_2$ given; counting efficiency (ca. 40%) calculated internally from their report; mass:epidermal peel area for C. cyanea (0.29 mg);[57] chl:mass for C. cyanea (0.41/mg chl/gm);[76] protein:chl ratio (107:1);[77] stomatal frequency (5386/cm²).[68]	Raschke & Dittrich, 1977[67]
Tulipa gesneriana	epidermal peels		Willmer et al., 1973[77]
Vicia faba	guard cells	See 1st entry under Outlaw et al., 1979.[52]	Outlaw et al., 1979[52]
Vicia faba	epidermal cells	See 1st entry under Outlaw et al., 1979.[52]	Outlaw et al., 1979[52]
Vicia faba	guard cells	See 1st entry under Outlaw et al., 1979.[52]	Outlaw et al., 1979[52]
Vicia faba	epidermal cells	See 1st entry under Outlaw et al., 1979.[52]	Outlaw et al., 1979[52]

increase, it was modest (as a percentage). This small change may be the source of the discrepancies among the reports. In addition to the published data,[81] we (unpublished) have conducted eight additional comparisons of sucrose in guard cells dissected from leaves with either open or closed stomata. In the three of these experiments where the stomatal aperture change was > +5 µm, the guard cell sucrose concentration increased significantly (Δ = +85 mmol/kg). In addition, circumstantial evidence for sucrose involvement in stomatal opening comes from the relatively

large percentage of ^{14}C sucrose synthesized by guard cell
protoplasts (our unpublished results in Table 5). I
speculate that sucrose may be an important osmotic component in the cytosol of open stomata. (It has been shown
to be important in osmoregulation in other plant
systems.[100,101]) Presumably, the bulk of potassium taken
up by guard cells would be sequestered in the vacuole (for
my reasons, see ref. [102]). This sequestration leaves open
the question of how cytosolic volume could be maintained
(i.e., if it is assumed that there is a negligible pres-
sure gradient across the tonoplast); import (or synthesis)
of sucrose could potentially decrease the cytosolic solute
potential sufficiently.

There is direct evidence that ^{14}C fixed in the
mesophyll is transported to the epidermis.[103] Vicia faba
leaves pulsed with $^{14}CO_2$ were quenched at various times
during $^{12}CO_2$ chase period. The leaves were freeze-
substituted and embedded under anhydrous conditions (to
prevent displacement of water-soluble compounds). ^{14}C in
paradermal sections of the epidermis increased with time
during the $^{12}CO_2$ chase period. These results indicate
carbon exchange between the mesophyll and epidermis. Two
groups[87,104] have reported that more ^{14}C accumulates in
epidermal peels if the leaf is labeled with $^{14}CO_2$ and then
the peel is taken (compared to $^{14}CO_2$ incubation with
detached epidermis). Though these last results could be
due to damage, they are consistent with ^{14}C transport to the
epidermis. In other work, Dittrich and Raschke[104] replaced
epidermal peels on pre-labeled mesophyll and reported ^{14}C
uptake by the peel. However, lack of tissue integrity
during this last work casts some doubt on the results.

There are data consistent with ^{14}C-sucrose transport
from the mesophyll to the epidermis.[105] Leaves of Vicia
were pulse-labeled with $^{14}CO_2$ for 1 min and epidermal peels
were taken at various times during the chase. Whole leaf
sucrose specific activity reach a maximum after 15 min,
that in the epidermis peaked much later (40 min). The most
direct interpretation is that most of the ^{14}C-sucrose found
in the epidermis was transported from the underlying
mesophyll. However, other interpretations are possible
(e.g., transport of precursors to ^{14}C-sucrose). It is
obvious that more work in this area of research is needed.

THE FATE OF MALATE (AND OTHER ORGANIC ANIONS) DURING STOMATAL CLOSURE

The reduction in guard cell malate concentration during stomatal closure could result from malate efflux or malate degradation by guard cells themselves. These are not exclusive options.

Van Kirk and Rashke[106] have shown that the reduction in epidermal strip malate concentration during stomatal closure was accompanied by malate appearance in the medium. The release was largest when stomata were closed by a treatment with abscisic acid. These findings are complemented by our observation[107] of high levels of malic enzyme in epidermal cells as well as in guard cells (Table 7). Other reports of NAD (Mn^{2+}) malic enzyme in epidermal peel extracts must be interpreted with caution. There are two sources of error (apparently not considered) that are particularly important in assaying for this enzyme in dilute extracts. The first source of error causes an underestimate and is due to reversal of the malic dehydrogenase equilibrium position during the course of the malic enzyme reaction.[109] The second source of error causes an over-estimate and is due to the Mn^{2+} stimulation of oxalacetate decarboxylation.[110] The latter would result in additional malate oxidation (to maintain the malic dehydrogenase equilibrium). Because the indicator for malate oxidative decarboxylation is the same as that for malate oxidation (i.e., NADH formation), the two are not ordinarily distinguished.

There are reports that carbons 1-3 of exogenously supplied malate are metabolized by a stomatal system to the starch in guard cells.[84,111,112] In two studies,[84,111] it was found that ^{14}C in C-4 of malate was not a donor for starch synthesis. However, these results need corroboration. One report[111] has been criticized[24] on the grounds that the [U-^{14}C]malate incubation was done at pH 3.3. It was argued[24] that this pH would result in death of all cells. In another of the reports,[84] the conversion of [U-^{14}C]malate to starch was low (678 Bq/mg protein); I calculate this to be less than 10^{-15} mol/guard cell pair·hr, which is not physiologically significant. From the other report,[112] I was unable to calculate precisely the rate of [U-^{14}C]malate conversion to starch, but I estimate that the

Table 7. Reported activities of malate-utilizing enzymes in epidermal tissues.

Enzyme	Reported Activity	% of "Mesophyll" on Basis Reported	(Values Calculated by Reviewer)		
			μmol/mg chl·hr	μmol/mg protein·hr	pmol/Guard Cell Pair·hr
(NADP⁺, Mg²⁺) Malic Enzyme	0.17 U/mg protein	708	(357)	(10.2)	(3.1)
	2550 μmol	2300	2550	(73)	(22)
	0.21 U/mg protein	875	--	(12.6)	--
	15.5 μmol/mg protein·hr	1192	(5661)	15.5	--
	2410 μmol/mg chl·hr	40200 (chl); 617 (protein)	2410	14.8	--
	1152 μmol/mg chl·hr	753	1152	(7.9)	--
	3360 μmol/mg chl·hr	19700 (chl); 194 (protein)	3360	3.1	--
	1364 μmol/mg chl·hr	1196	1364	--	--
(NADP⁺, Mn²⁺) Malic Enzyme	45.8 μmol/mg protein·hr	2082	(16425)	45.8	--
(NAD⁺, Mg²⁺) Malic Enzyme	0.1 μmol/mg protein·hr	100	(36)	0.1	--
(NAD⁺, Mn²⁺) Malic Enzyme	0.66 U/mg protein	880	(1416)	(40)	(12)
	0.27 U/mg protein	360	--	(16.2)	--
	2.4 μmol/mg protein·hr	400	(876)	2.4	--
(NAD⁺) Malic Dehydrogenase	580 μmol/mg chl·hr	--	580	(16.5)	(5.0)
	182 μmol/mg protein·hr	95	(66430)	182	--
	≥ 8360 μmol/mg chl·hr	1091 (chl); 73 (protein)	≥ 8360	≥ 64	--
	1968 μmol/mg chl·hr	299	1968	(13.6)	--
	309,000 μmol/mg chl·hr	17965 (chl); 79 (protein)	309,000	177	--
	2562 μmol/mg chl·hr	319	2562	--	--
(NADP⁺) Malic Dehydrogenase	10.6 μmol/mg protein·hr	461	(3869)	10.6	--
	≥ 2160 μmol/mg chl·hr	2769 (chl); 55 (protein)	≥ 2160	≥ 15.8	--
	854 μmol/mg chl·hr	694	854	(5.9)	--
	1150 μmol/mg chl·hr	4107 (chl); 40 protein)	1150	0.96	--
	1258 μmol/mg/chl·hr	912	1258	--	--

Species	Tissue System	Comments	Reference
Vicia faba	guard cells	Protein:chl ratio (35:1);[47] protein content 5.2%;[49] guard cell pair mass (6 ng).[44]	Outlaw et al., 1981[107]
Vicia faba	guard cell protoplasts	See 1st entry under Outlaw et al., 1981[107]	Schnabl, 1981[80]
Vicia faba	epidermal cells	Protein content (7.1%).[49]	Outlaw et al., 1981[107]
Commelina cyanea	epidermal peels	Protein:chl ratio (365:1) derived from their text.	Thorpe et al., 1978[76]
Commelina communis	epidermal peels		Willmer et al., 1973[77]
Commelina benghalensis	epidermal peels	Value in parentheses calculated by Thorpe et al., 1978[76]	Rama Das & Raghavendra, 1974[79]
Tulipa gesneriana	epidermal peels		Willmer et al., 1973[77]
Tridax procumbens	epidermal peels		Rama Das & Raghavendra, 1974[79]
Commelina cyanea	epidermal peels	See 1st entry under Thorpe et al., 1978.[76]	Thorpe et al., 1978[76]
Commelina cyanea	epidermal peels	See 1st entry under Thorpe et al., 1978.[76]	Thorpe et al., 1978[76]
Vicia faba	guard cells	See 1st entry under Outlaw et al., 1981.[107]	Outlaw et al., 1981[107]
Vicia faba	epidermal cells	See 1st entry under Outlaw et al., 1981.[107]	Outlaw et al., 1981[107]
Commelina cyanea	epidermal peels	See 1st entry under Thorpe et al., 1978.[76]	Thorpe et al., 1978[76]
Vicia faba	guard cell protoplasts	See 1st entry under Outlaw et al., 1981[107]	Schnabl, 1981[80]
Commelina cyanea	epidermal peels	See 1st entry under Thorpe et al., 1978.[76]	Thorpe et al., 1978[76]
Commelina communis	epidermal peels		Willmer et al., 1973[77]
Commelina benghalensis	epidermal peels	Value in parentheses calculated by Thorpe et al., 1978[76]	Rama Das & Raghavendra, 1974[79]
Tulipa gesneriana	epidermal peels		Willmer et al., 1973[77]
Tridax procumbens	epidermal peels		Rama Das & Raghavendra, 1974[79]
Commelina cyanea	epidermal peels	See 1st entry under Thorpe et al., 1978.[76]	Thorpe et al., 1978[76]
Commelina communis	epidermal peels		Willmer et al., 1973[77]
Commelina benghalensis	epidermal peels	Value in parentheses calculated by Thorpe et al., 1978[76]	Rama Das & Raghavendra, 1974[79]
Tulip gesneriana	epidermal peels		Willmer et al., 1973[77]
Tridax procumbens	epidermal peels		Rama Das & Raghavendra, 1974[79]

Table 8. Reported activities of phosphoenolpyruvate-synthesizing enzymes in epidermal tissue.

Enzyme	Reported Activity	% of "Mesophyll" on Basis Reported	(Values Calculated by Reviewer)		
			μmol/mg chl·hr	μmol/mg protein·hr	pmol/Guard Cell Pair·hr
Phosphoenol-pyruvate carboxykinase	N.D. (<70 mmol/kg·hr)	N.D./N.D.	(<47)	(<1.3)	(<0.4)
	N.D.	N.D./--	N.D.	(N.D.)	(N.D.)
	N.D.	N.D./N.D.	--	--	--
Pyruvate, ortho-phosphate, dikinase	N.D. (<20 mmol/kg·hr)	N.D./N.D.	(<14)	(<0.4)	(<0.1)
	21	81	21	(0.6)	(0.2)
	N.D.	N.D./N.D.	--	--	--
	342 μmol/mg chl·hr	1052	342	(2.4)	--
	768 μmol/mg chl·hr	1896	768	--	--

upper limit was also less than 10^{-15} mol/guard cell pair·hr. I conclude that qualitatively the results are consistent with the malate's being decarboxylated to pyruvate, which then is converted to starch. However, quantitatively, the results to date are insufficient to support this conclusion.

Additional evidence that malate is metabolized to pyruvate are the high levels of malic enzyme and non-detectible levels of phosphoenolpyruvate carboxykinase (Table 8) in extracts of stomatal systems. Because at least some of the 3-C fragment is metabolized to starch,[84,111,112] it must be phosphorylated to phospho-enolpyruvate (for entry into gluconeogenesis). How this occurs is not known. This step might be catalyzed by pyruvate, orthophosphate dikinase, but two groups[76,107] have been unable to detect this enzyme in epidermal tissue. There are also two positive reports.[79,80] I have reserva-tions about interpreting one of these reports[79] because of the enzyme coupling methods used (cf. ref. 107). The methods used for the other report[80] appear sufficient. However, the result showing 26 μmol/mg chl·hr for mesophyll of a C_3 plant does not conform to previous surveys (e.g., ref. 108). The general question of malate metabolism to starch by guard cells is an open one. It deserves attention.

Species	Tissue System	Comments	Reference
Vicia faba	guard cells	Protein:chl ratio (35:1);[47] protein content (5.2%);[49] guard cell pair mass (6 ng).[44]	Outlaw et al., 1981[107]
Vicia faba	guard cell protoplasts		Schnabl, 1981[80]
Commelina cyanea	epidermal peels		Thorpe et al., 1978[76]
Vicia faba	guard cells	See 1st entry under Outlaw et al., 1981.[107]	Outlaw et al., 1981[107]
Vicia faba	guard cell protoplasts	See 1st entry under Outlaw et al., 1981.[107]	Schnabl, 1981[80]
Commelina cyanea	epidermal peels		Thorpe et al., 1978[76]
Commelina benghalensis	epidermal peels	Value in parentheses calculated by Thorpe et al., 1978.[76]	Rama Das & Raghavendra, 1974[79]
Tridax procumbens	epidermal peels		Rama Das & Raghavendra, 1974[79]

LOOSE ENDS

There are three aspects of carbon metabolism in guard cells that I have not addressed. These omissions are due either to lack of sufficient data or to my inability to make an interpretation of the conflicting reports. For the role of ATP hydrolysis in ion uptake in guard cells, I refer the reader to general reviews[113,114] and to recent papers[115-117] for an introduction to the literature. Despite a huge body of literature on abscisic acid and stomata, direct proof for its role in stomatal aperture size regulation in situ is lacking; neither abscisic acid nor any enzymes specific for it has been quantified in guard cells. (However, a report is forthcoming on its measurement in guard cell protoplasts.[116]) I refer the reader to a recent general review[118] and to a recent report on epidermal tissue[119] for an introduction to that literature. Whether both photosystems are present in guard cells remains controversial; the reader is referred to refs. 47, 98, 120, and 121 for an introduction to that literature.

SUMMARY

I tentatively conclude from this "inventory" of the current literature that the biochemical outline presented

in Fig. 1 is correct. Stomata open as guard cells take up
potassium. Cytoplasmic pH is stabilized by Cl^- uptake and
the synthesis of malate (or other organic anions), the
carbon skeleton of which is derived from starch. The PCRP
is absent. Sucrose is suggested to reduce the cytosolic
solute potential sufficiently. During stomatal closure,
malate is released from guard cells or is metabolized back
to starch.

ACKNOWLEDGEMENTS

I thank Dr. A. Thistle for manuscript revisions and
Drs. M. Bodson, J. Croxdale, R. Hampp and T. Roberts for
manuscripts suggestions. I am especially grateful to Drs.
W. Allaway, K. Raschke, H. Schnabl and C. Willmer for
critical reviews. The support by NSF and DOE of work in
my laboratory during the preparation of this review is
appreciated.

REFERENCES

1. Allaway, W. G. 1976. Influence of stomatal behavior
    on long distance transport. In Transport and
    transfer processes (I. F. Wardlaw, J. B. Passioura,
    eds). Academic Press, New York, pp. 295-311.
2. Barrs, H. D. 1971. Cyclic variations in stomatal
    aperture, transpiration and leaf water potential
    under constant environmental conditions. Annu. Rev.
    Plant Physiol. 22: 223-236.
3. Cowan, I. R. 1977. Stomatal behavior and environment.
    Adv. Bot. Res. 4: 117-228.
4. Hsiao, T. C. 1974. Plant responses to water stress.
    Annu. Rev. Plant Physiol. 24: 519-570.
5. Hsiao, T. C. 1976. Stomatal ion transport. In Trans-
    port in plants II (U. Lüttge, M. G. Pittman, eds.).
    Encyclopedia of Plant Physiology, New Series.
    Springer-Verlag, Berlin, pp. 195-221.
6. MacRobbie, E. A. C. 1977. Functions of ion transport
    in plant cells and tissues. In Plant Biochemistry
    II (D. H. Northcote, ed), International Review of
    Biochemistry, Vol. 13. University Park Press,
    Baltimore, pp. 226-234.
7. Meidner, H., C. M. Willmer. 1975. Mechanics and
    metabolism of guard cells. Curr. Adv. Plant Sci.
    17: 1-15.

8. Milthorpe, F. L.   1970.   The significance and mechanism
    of stomatal movement.   Aust. J. Sci. 32: 31-35.
9. Milthorpe, F. L., C. J. Pearson, S. Thrower.   1974.
    The metabolism of guard cells.   In Mechanisms of
    regulation of plant growth (R. L. Bieleski, A. R.
    Ferguson, M. M. Cresswell, eds), Bull. 12.   The
    Royal Society of New Zealand, Wellington,
    pp. 439-443.
10. Palevitz, B. A.   198X.   Stomatal complexes as a model
    of cytoskeleton participation in cell differentia-
    tion.   In The cytoskeleton in plant growth and
    development (C. Lloyd, ed).   Academic Press, New
    York, in press.
11. Pospíšilova, J., J. Solárová.   1980.   Environmental and
    biological control of diffusive conductances of
    adaxial and abaxial leaf epidermes.
    Photosynthetica 14: 90 127.
12. Raschke, K.   1975.   Stomatal action.   Annu. Rev. Plant
    Physiol. 26: 309-340.
13. Raschke, K.   1976.   How stomata resolve the dilemma of
    opposing priorities.   Phil. Trans. R. Soc. Lond. B.
    273: 551-560.
14. Raschke, K.   1976.   Transfer of ions and products of
    photosynthesis to guard cells.   In Transport and
    transfer processes in plants (I. F. Wardlaw and
    J. B. Passioura, eds).   Academic Press, New York,
    pp. 203-215.
15. Raschke, K.   1977.   The stomatal turgor mechanism and
    its response to CO_2 and abscisic acid: observations
    and a hypothesis.   In Regulation of cell membrane
    activites in plants (E. Marré, O. Ciferri, eds).
    Elsevier/North-Holland Biomedical Press, Amsterdam,
    pp. 173-183.
16. Raschke, K.   1979.   Movements of stomata.   In
    Physiology of movements (W. Haupt, M. E. Feinleib,
    eds.).   Encyclopedia of Plant Physiology, Vol. 7.
    Springer-Verlag, Berlin, pp. 383-441.
17. Zeiger, E., A. J. Bloom, P. K. Hepler.   1978.   Ion
    transport in stomatal guard cells: a chemiosmotic
    hypothesis.   What's New in Plant Physiology 9: 29-32.
18. Jarvis, P. G., T. A. Mansfield (eds.).   198X.
    Stomatal physiology.   Cambridge University Press,
    Cambridge, in press.

19. Rogers, C. A. 1980. Integrated view of guard cells.
    Symposium Series of the Southern Section of the
    American Society of Plant Physiologists. Houston,
    pp. 1-64.

20. Sen, D. N., D. D. Chawan, R. P. Bansal (eds). 1979.
    Structure, function and ecology of stomata. Bishen
    Singh Mahendra Pal Singh, Dehra Dun (India).

21. Willmer, C. M. 198X. Stomata. Longman, London.
    In Press.

22. Meidner, H., T. A. Mansfield. 1968. Physiology of
    Stomata. McGraw-Hill, New York.

23. Outlaw, W. H., Jr. 1980. Unique aspects of carbon
    metabolism in guard cells of Vicia faba L. In
    Integrated view of guard cells (C. A. Rogers, ed).
    Symposium Series of the Southern Section of the
    American Society of Plant Physiologists, Houston,
    pp. 4-18.

24. Willmer, C. M. 1981. Guard cell metabolism. In
    Stomatal Physiology (P. G. Jarvis, T. A. Mansfield,
    eds). Cambridge University Press, Cambridge.
    pp. 87-102.

25. Yemm, E. W., A. J. Willis. 1954. Stomatal movements
    and changes of carbohydrate in leaves of
    Chrysanthemum maximum. New Phytol. 53: 373-397.

26. Fischer, R. A. 1968. Stomatal opening: role of
    potassium uptake by guard cells. Science
    160: 784-785.

27. Fischer, R. A. 1968. Stomatal opening in isolated
    epidermal strips of Vicia faba. I. Response to
    light and CO$_2$-free air. Plant Physiol.
    43: 1947-1952.

28. Fischer, R. A., T. C. Hsiao. 1968. Stomatal opening
    in isolated epidermal strips of Vicia faba. II.
    Response to KCl concentration and the role of
    potassium absorption. Plant Physiol. 43: 1953-1958.

29. Fujino, M. 1967. Adenosinetriphosphate and adenosine-
    triposphatase in stomatal movement. Sci. Bull. Fac.
    Educ. (Nagaski University) 18: 1-74.

30. Imamura, S. 1943. Research about the mechanism of
    the turgor-fluctuation of the stomatal guard cells
    (in German). Jap. J. Bot. 12: 251-346.

31. Raschke, K., G. D. Humble. 1973. No uptake of anions
    required by opening stomata of Vicia faba: guard
    cells release hydrogen ions. Planta 115: 47-57.

32. Humble, G. D., K. Raschke. 1971. Stomatal opening quantitatively related to potassium transport. Evidence from electron probe analysis. Plant Physiol. 48: 447-453.

33. Raschke, K., M. P. Fellows. 1971. Stomatal movement in Zea mays: shuttle of potassium and choride between guard cells and subsidiary cells. Planta 101: 296-316.

34. Pallaghy, C. K., R. A. Fisher. 1974. Metabolic aspects of stomatal opening and ion accumulation by guard cells in Vicia faba. Z. Pflanzenphysiol. 71: 332-334.

35. Penny, M. G., L. S. Kelday, D. J. F. Bowling. 1976. Active choride transport in the leaf epidermis of Commelina communis in relation to stomatal activity. Planta 130: 291-294.

36. Schnabl, H., K. Raschke. 1980. Potassium chloride as stomatal osmoticum in Allium cepa L., a species devoid of starch in guard cells. Plant Physiol. 63: 88-93.

37. Clarkson D. T. 1974. Ion transport and cell structure in plants. McGraw-Hill Ltd., Berkshire.

38. Jacoby, B., G. G. Laties. 1971. Bicarbonate fixation and malate compartmentation in relation to salt-induced stoichiometric synthesis of organic acid. Plant Physiol. 47: 525-531.

39. Smith, F. A., J. A. Raven. 1979. Intracellular pH and its regulation. Annu. Rev. Plant Physiol. 30: 289-311.

40. Davies, D. D. 1979. The central role of phosphoenol-pyruvate in plant metabolism. Annu. Rev. Plant Physiol. 30: 131-158.

41. Sawhney, B. L., I. Zelitch. 1969. Direct determination of potassium ion accumulation in guard cells in relation to stomatal opening in light. Plant Physiol. 44: 1350-1354.

42. Penny, M. G., D. J. F. Bowling. 1974. A study of potassium gradients in the epidermis of intact leaves of Commelina communis in relation to stomatal opening. Planta 122: 209-212.

43. Allaway, W. G., T. C. Hsiao. 1973. Preparation of rolled epidermis of Vicia faba L. so that stomata are the only viable cells: analysis of guard cell potassium by flame photometry. Aust. J. Biol. Sci. 26: 309-318.

44. Outlaw, W. H., Jr., O. H. Lowry. 1977. Organic acid and potassium accumulation in guard cells during stomatal opening. Proc. Nat. Acad. Sci. USA 74: 4434-4438.

45. Raschke, K., H. Schnabl. 1978. Availability of chloride affects the balance between potassium chloride and potassium malate in guard cells of Vicia faba L. Plant Physiol. 62: 84-87.

46. Outlaw, W. H., Jr., C. L. Schmuck, N. E. Tolbert. 1976. Photosynthetic carbon metabolism in the palisade parenchyma and spongy parenchyma of Vicia faba L. Plant Physiol. 58: 186-189.

47. Outlaw, W. H., Jr., B. C. Mayne, V. E. Zenger, J. Manchester. 1981. Presence of both photosystems in guard cells of Vicia faba L. Implications for environmental signal processing. Plant Physiol. 67: 12-16.

48. Jones, M. G. K., W. H. Outlaw, Jr., O. H. Lowry. 1977. Enzymic assay of 10^{-7} to 10^{-14} moles of sucrose in plant tissues. Plant Physiol. 60: 379-383.

49. Outlaw, W. H., Jr., J. Manchester, V. E. Zenger. 1981. The relationship between protein content and dry weight of guard cells and other single cell samples of Vicia faba L. leaflet. Histochem. J. 13: 329-336.

50. Paul, J. S., J. A. Bassham. 1977. Maintenance of high photosynthetic rates in mesophyll cells iso-lated from Papaver somniferum L. Plant Physiol. 60:775-778.

51. Hammel, K. E., K. L. Cornwell, J. A. Bassham. 1979. Stimulation of dark CO_2 fixation by ammonia in isolated mesophyll cells of Papaver somniferum L. Plant Cell Physiol. 20: 1523-1529.

52. Outlaw, W. H., Jr., J. Manchester, C. A. DiCamelli, D. D. Randall, B. Rapp, G. M. Veith. 1979. Photo-synthetic carbon reduction pathway is absent in chloroplasts of Vicia faba guard cells. Proc. Nat. Acad. Sci. USA 76: 6371-6375.

53. Outlaw, W. H., Jr., J. Kennedy. 1978. Enzymic and substrate basis for the anaplerotic step in guard cells. Plant Physiol. 62: 648-652.

54. Outlaw, W. H., Jr., J. Manchester, C. A. DiCamelli. 1979. Histochemical approach to properties of Vicia faba guard cell phosphoenolpyruvate carboxy-lase. Plant Physiol. 64: 269-272.

55. Allaway, W. G. 1981. Anions in stomatal regulation.
    In Stomatal physiology (P. G. Jarvis, T. A.
    Mansfield, eds.). Cambridge University Press,
    Cambridge, pp. 71-85

56. Allaway, W. G. 1973. Accumulation of malate in guard
    cells of Vicia faba during stomatal opening.
    Planta 110: 63-70.

57. Pearson, C. J. 1973. Daily changes in stomatal
    aperture and in carbohydrates and malate within
    epidermis and mesophyll of leaves of Commelina
    cyanea and Vicia faba. Aust. J. Biol. Sci.
    26: 1035-1044.

58. Pallas, J. E., Jr., B. G. Wright. 1973. Organic acid
    changes in the epidermis of Vicia faba and their
    implication in stomatal movement. Plant Physiol.
    51: 588-590.

59. Fischer, R. A. 1971. Role of potassium in stomatal
    opening in the leaf of Vicia faba. Plant Physiol.
    47: 555-558.

60. Ogawa, T., H. Ishikawa, K. Shimada, K. Shibata. 1978.
    Synergistic action of red and blue light and action
    spectra of malate formation in guard cells of Vicia
    faba L. Planta 142: 61-65.

61. Van Kirk, C. A., K. Raschke. 1978. Presence of
    chloride reduces malate production in epidermis
    during stomatal opening. Plant Physiol.
    61: 361-364.

62. Schnabl, H. 1978. The effect of Cl^- upon the sensi-
    tivity of starch-containing and starch-deficient
    stomata and guard cell protoplasts toward potassium
    ions, fusicoccin and abscisic acid. Planta
    14: 95-100.

63. Schnabl, H. 1980. Anion metabolism as correlated
    with volume changes of guard cell protoplasts. Z.
    Naturforsch. 35: 621-626.

64. Pearson, C. J., F. L. Milthorpe. 1974. Structure,
    carbon dioxide fixation, and metabolism of stomata.
    Aust. J. Plant Physiol. 1: 221-236.

65. Bowling, D. J. F. 1976. Malate-switch hypothesis to
    explain the action of stomata. Nature (London)
    262: 393-394.

66. Travis, A. J., T. A. Mansfield. 1977. Studies of
    malate formation in 'isolated' guard cells. New
    Phytol. 78: 541-546.

67. Raschke, K., P. Dittrich. 1977. [^{14}C]carbon dioxide
     fixation by isolated leaf epidermis with stomata
     closed or open. Planta 134: 69-75.
68. Willmer, C., R. Kanai, J. E. Pallas, Jr., C. C. Black,
     Jr. 1973. Detection of high levels of phosphoenol-
     pyruvate carboxylase in leaf epidermal tissues and
     its significance in stomatal movements. Life Sci.
     12: 151-155.
69. Squire, G. R., T. A. Mansfield. 1972. A simple method
     of isolating stomata on detached epidermis by low
     pH treatment: observations of the importance of
     subsidiary cells. New Phytol. 71: 1033-1043.
70. Rutter, J. C., W. R. Johnston, C. M. Willmer. 1977.
     Free sugars and organic acids in the leaves of
     various plant species and their compartmentation
     between tissues. J. Exptl. Bot. 28: 1019-1028.
71. Contour-Ansel, D., P. Longuet. 1979. Comparaison
     entre les teneurs en acides organiques d'épidermis
     foliaires isolés de Pelargonium X hortorum, en
     relation avec l'état d'ouverture ou de fermeture
     des stomates: l'acide malique joue-t-il toujours
     un rôle déterminant dans les mouvements stomatiques?
     Physiol. Vég. 17: 337-346.
72. Lowry, O. H., J. V. Passonneau. 1972. A flexible
     system of enzymatic analysis. Academic Press,
     New York.
73. Outlaw, W. H., Jr. 1978. Biochemical analysis of
     single plant cells. What's New in Plant Physiology
     9: 21-24.
74. Outlaw, W. H., Jr. 1980. A descriptive evaluation of
     quantitative histochemical methods based on
     pyridine nucleotides. Annu. Rev. Plant Physiol.
     31: 299-311.
75. Pallas, J. E., Jr., R. A. Dilley. 1972. Photophos-
     phorylation can provide sufficient adenosine 5'-
     triphosphate to drive K$^+$ movements during stomatal
     opening. Plant Physiol. 49: 649-650.
76. Thorpe, N., C. J. Brady, F. L. Milthorpe. 1978.
     Stomatal metabolism: Primary carboxylation and
     enzyme activities. Aust. J. Plant Physiol.
     5:485-493.
77. Willmer, C. M., J. E. Pallas, Jr., C. C. Black, Jr.
     1973. Carbon dioxide metabolism in leaf epidermal
     tissue. Plant Physiol. 52: 448-452.

78. Donkin, M., E. S. Martin. 1980. Studies on the properties of carboxylating enzymes in the epidermis of Commelina communis. J. Exptl. Bot. 31: 357-363.
79. Rama Das, V. S., A. S. Raghavendra. 1974. Control of stomatal opening by pyruvate metabolism in light. Ind. J. Exptl. Biol. 12: 425-428.
80. Schnabl, H. 1981. The compartmentation of carboxylating and decarboxylating enzymes in guard cell protoplasts. Planta 152: 307-313.
81. Outlaw, W. H., Jr., J. Manchester. 1979. Guard cell starch concentration quantitatively related to stomatal aperture. Plant Physiol. 64: 79-82.
82. Mouravieff, I. 1972. Microphotométrie des fluctuations de la teneur en amidon des stomates en rapport avec l'ouverture de l'ostiole a la lumière, en présence ou en absence de gaz carbonique. Annales des Sciences Naturelles, Botanique 1: 361-368.
83. Outlaw, W. H., Jr., D. B. Fisher. 1975. Compartmentation in Vicia faba leaves. III. Photosynthesis in the spongy and palisade parenchyma. Aust. J. Plant Physiol. 2: 435-439.
84. Schnabl, H. 1980. CO_2 and malate metabolism in starch-containing and starch-lacking guard cell protoplasts. Planta 149: 52-58.
85. Schnabl, H. 1977. Isolation and identification of soluble polysaccharides in epidermal tissue of Allium cepa. Planta 135:307-311.
86. Willmer, C. M., P. Dittrich. 1974. Carbon dioxide fixation by epidermal and mesophyll tissues of Tulipa and Commelina. Planta 117: 123-132.
87. Willmer, C. M., N. Thorpe, J. C. Rutter, F. L. Milthorpe. 1978. Stomatal mechanism: Carbon dioxide fixation in attached and detached epidermis of Commelina. Aust. J. Plant Physiol. 5: 767-778.
88. Zelitch, I., D. A. Walker. 1964. The role of glycolic acid metabolism in opening of leaf stomata. Plant Physiol. 39:856-862.
89. Zelitch, I. 1961. Biochemical control of stomatal opening of leaves. Proc. Natl. Acad. Sci. USA 47: 1423-1433.
90. Kaiser, W. M., J. A. Bassham. 1979. Light-dark regulation of starch metabolism in chloroplasts. I. Levels of metabolites in chloroplasts and medium during light-dark transition. Plant Physiol. 63: 105-108.

91. Preiss, J. 1973. Adenosine diphosphoryl glucose pyrophosphorylase. In The enzymes (P. Boyer, ed). Academic Press, New York. 8: 73-119.

92. Schnabl, H. 1980. Rapid gluconeogenesis in starch-containing guard cell protoplasts. In Plant membrane transport: current conceptual issues (R. M. Spanswick, W. J. Lucas, J. Dainty, eds). Elsevier/North Holland Biomedical Press, Amsterdam, pp. 455-456.

93. Lorimer, G. H. 1981. The carboxylation and oxygenation of ribulose-1,5-bisphosphate: the primary event in photosynthesis and photorespiration. Annu. Rev. Plant Physiol. 32: 349-383.

94. Godavari, H. R., S. S. Badour, E. R. Waygood. 173. Isocitrate lyase in green leaves. Plant Physiol. 51: 863-867.

95. Zelitch, I. 1973. Alternate pathways of glycollate synthesis in tobacco and maize leave in relation to rates of photorespiration. Plant Physiol. 51: 299-305.

96. Randall, D. D., N. E. Tolbert. 1971. 3-phospho-glycerate in plants. J. Biol. Chem. 17: 5510-5517.

97. Thorpe, N., C. M. Willmer, F. L. Milthorpe. 1979. Stomatal metabolism: carbon dioxide fixation and labeling patterns during stomatal movement in Commelina cyanea. Aust. J. Plant Physiol. 6: 409-416.

98. Zeiger, E., P. Armond, A. Melis. 1981. Fluorescence properties of guard cell chloroplasts. Evidence for linear electron transport and light-harvesting pigments of photosystems I and II. Plant Physiol. 67: 17-20.

99. Madhavan, S., B. N. Smith. 1982. Localization of ribulose bisphosphate carboxylase in guard cells by an indirect immunofluorescence technique. Plant Physiol. In Press.

100. Kirst, G. O. 1980. $^{14}CO_2$ fixation in Valonia utri-cularis subjected to osmotic stress. Plant Sci. Lett. 18: 155-160.

101. Jones, M. M., C. B. Osmond, N. C. Turner. 1980. Accumulation of solutes in leaves of sorghum and sunflower in response to water deficits. Aust. J. Plant Physiol. 7: 193-205.

102. Zimmerman, U. 1978. Physics of turgor- and osmo-regulation. Annu. Rev. Plant Physiol. 29: 121-148.

103. Outlaw, W. H., Jr., D. B. Fisher. 1975. Compartmenta-
     tion in Vicia faba leaves. I. Kinetics of ^{14}C in
     the tissues following pulse labeling. Plant
     Physiol. 55: 699-703.
104. Dittrich, P., K. Raschke. 1977. Uptake and metabo-
     lism of carbohydrate by epidermal tissue. Planta
     134: 83-90.
105. Outlaw, W. H., Jr., D. B. Fisher, A. L. Christy. 1975.
     Compartmentation in Vicia faba leaves. II. Kinetics
     of ^{14}C-sucrose redistribution among individual
     tissues following pulse labeling. Plant Physiol.
     55: 704-711.
106. Van Kirk, C. A., K. Raschke. 1978. Release of malate
     from epidermal strips during stomatal closure.
     Plant Physiol. 61: 374-375.
107. Outlaw, W. H., Jr., J. Manchester, P. H. Brown. 1981.
     High levels of malic enzyme activities in Vicia
     faba L. epidermal tissue. Plant Physiol.
     68: 1047-1051.
108. Hatch, M. D., C. R. Slack. 1968. A new enzyme for the
     interconversion of pyruvate and phosphopyruvate and
     its role in the C_4 dicarboxylic acid pathway of
     photosynthesis. Biochem. J. 106: 141-146.
109. Outlaw, W. H., Jr., J. Manchester. 1980. Conceptual
     error in determination of NAD -malic enzyme in
     extracts containing NAD -malic dehydrogenase.
     Plant Physiol. 65: 1136-1138.
110. Mazelis, M., B. Vennesland. 1957. Carbon dioxide
     fixation into oxalacetate by higher plants.
     Plant Physiol. 32: 591-600.
111. Dittrich, P., K. Raschke. 1977. Malate metabolism in
     isolated epidermis of Commelina communis L. in rela-
     tion to stomatal functioning. Planta 134: 77-81.
112. Willmer, C. M., J. C. Rutter. 1977. Guard cell malic
     acid metabolism during stomatal movements. Nature
     (London) 269: 327-328.
113. Spanswick, R. M., W. J. Lucas, J. Dainty. 1980. Plant
     Membrane Transport: Current Conceptual Issues.
     Elsevier, Amsterdam.
114. Poole, R. J. 1978. Energy coupling for membrane
     transport. Annu. Rev. Plant Physiol. 29:437-460.
115. Kasamo, K. 1979. Characterization of membrane-bound
     Mg^{2+}-activated ATPase isolated from the lower
     epidermis of tobacco leaves. Plant Cell Physiol.
     20: 281-292.

116. Weiler, E. W., H. Schnabl, C. Hornberg. 198X. Stress-
     related levels of abscisic acid in guard cell proto-
     plasts of Vicia faba L. Planta, submitted.
117. Lurie, S., D. L. Hendrix. Differential ion stimula-
     tion of plasmalemma adenosine triphosphatase from
     leaf epidermis and mesophyll of Nicotiana rustica L.
     Plant Physiol. 63:936-939.
118. Walton, D. C. 1980. Biochemistry and physiology of
     abscisic acid. Annu. Rev. Plant Physiol.
     31: 453-489.
119. Singh, B. N., E. Galson, W. Dashek, D. C. Walton.
     1979. Abscisic acid levels and metabolism in leaf
     epidermal tissue of Tulipa gesneriana L. and
     Commelina communis L. Planta 146: 135-138.
120. Melis, A., E. Zeiger. 198X. Fluorescence transients
     in mesophyll and guard cell chloroplasts. Evidence
     for CO_2 modulation of photophosphorylation in guard
     cells. Plant Physiol., submitted.
121. Schnabl, H., R. Hampp. 1980. Vicia guard cell proto-
     plasts lack photosystem II activity.
     Naturwissenschaften 67: 465-466.

Chapter Seven

CELLULAR ASPECTS OF C$_4$ LEAF METABOLISM

W. H. CAMPBELL

Chemistry Department
College of Environmental Science and Forestry
State University of New York
Syracuse, NY   13210

C. C. BLACK
Biochemistry Department
University of Georgia
Athens, GA   30602

INTRODUCTION

The discovery of C$_4$ photosynthesis heralded the beginning of a new understanding of cellular specialization among the leaf cells of higher plants. The essence of C$_4$ photosynthesis and of this new understanding of leaf metabolism can be stated simply: two green cell types are always found in a C$_4$ plant; and these cells do not have a fixed morphological or anatomical arrangement but share, in an interdependent manner, the tasks of photosynthetic energy capture, storage, and utilization. By taking this view of the C$_4$ plant, we will describe photosynthetic metabolism in C$_4$ leaf cells as a cooperative activity between the mesophyll cells, which trap atmospheric CO$_2$ as readily transportable (but nevertheless reactive) four-carbon acids, and the bundle sheath cells, which decarboxylate the acids and refix the CO$_2$ via the C$_3$ cycle. This character of interdependence among leaf cells in which cell type has

223

several required roles to complete a metabolic cycle is not a characteristic of C_3 photosynthesis plants. Among the cells of the C_3 leaf, the anatomy of individual photosynthetic cells may vary (e.g., spongy versus palisade mesophyll cells), but each cell type appears to serve the leaf as a semi-autonomous unit of the same metabolic type without a great deal of metabolic interdependence.

This presentation is intended to review findings on the shared activities of mesophyll and bundle sheath cells typical of C_4 plants and is not a general review of C_4 plants which has been presented.[1] A variety of other features such as environmental physiology, growth rates, yields, and plant competition are essential for a complete understanding of C_4 plants but rather than review this information, we will integrate these ideas and data with metabolic studies and cite key references for interested readers. We will focus on the leaf and cell anatomy associated with C_4 photosynthesis and place emphasis upon the biochemistry of carbon, nitrogen, and sulfur metabolism in mesophyll and bundle sheath cells.

ANATOMY OF C_4 LEAVES AND CHLOROPLASTS

One of the most clear-cut and interesting features of C_4 photosynthesis is the requirement of a physical relationship between the cooperating cells of the C_4 leaf. A typical C_4 plant leaf when viewed in cross-section has Kranz leaf anatomy (Figure 1). Kranz leaf anatomy was described over one hundred years ago,[2] but it was not known that the radially arranged layers of bundle sheath and mesophyll cells were of different metabolic natures and acted together to complete the process of photosynthesis. Following the discovery of C_4 photosynthetic CO_2 fixation,[3,4] it was realized simultaneously by four groups that C_4 plants possess Kranz leaf anatomy,[5-8] which had been discovered earlier.[2] As the metabolic cooperation of C_4 green cells became understood, it was established that Kranz leaf anatomy is an essential component of the C_4 photosynthetic system. Today we strongly associate Kranz leaf anatomy with C_4 plants and all known plants with green bundle sheath cells surrounded by a single layer of green mesophyll cells are C_4 plants.[1] Thus, our concept of cellular specialization in photosynthesis had its origin in early studies of gross leaf anatomy with higher plants.

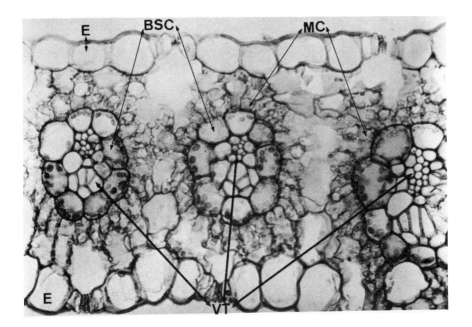

Figure 1. Light micrograph of a <u>Sorghum</u> <u>bicolor</u> leaf cross-
section showing Kranz leaf anatomy with vascular tissues
(VT) surrounded by bundle sheath cells (BSC) then mesophyll
cells (MC). E = epidermis.

However, not all C_4 plants have Kranz leaf anatomy
and not all plants with green bundle sheath cells are C_4
plants.[1,9-11] Despite these caveats, all plants that can
be clearly characterized as a C_4 photosynthetic type have
two types of cells that cooperate in net CO_2 fixtation
(Table 1). For example, <u>Suaeda</u> <u>monoica</u>, which has only
rudimentary bundle sheath cell chloroplasts, has two layers
of green cells under the epidermis of its semi-cylindrical
leaves.[11] In sharp contrast to <u>Suaeda</u>, <u>Panicum</u> <u>milioides</u>,
one of the so called C_3/C_4 hybrids,[10] has well developed
bundle sheath chloroplasts, but lacks the proper enzyme
balance to perform as C_4 photosynthetic cell types
(Table 1). The mere presence of two types of green cells
is insufficient data to establish the presence of C_4 photosynthesis.

Table 1. Some variations in C_4 anatomy in green tissue cross sections and a conclusion.

I. The vast majority of C_4 plants have Kranz leaf anatomy. Figure 1.

II. Sedges have a non-green layer of cells between the green bundle sheath and mesophyll cell layers.

III. Astridia has single and double layers of green bundle sheaths.

IV. Suaeda monoica has a succulent, semi-cylindrical leaf with two layers of green cells beneath the epidermis, but no green bundle sheaths.

V. Salsola kali has a succulent, cylindrical, leaf without grean bundle sheaths. The chlorenchyma is differentiated into a green palisade and a chlorenchymatous sheath.

VI. In Portulaca grandifloria, mesophyll cells only occur between the epidermis and the the bundle sheath cells.

VII. Other specialized anatomical modifications exist in specific plants.

Conclusion: In all cases of documented C_4 metabolism two green cell types are present!

A light microscopic examination of a leaf (as in Figure 1) will allow a tentative identification of a plant as a C_4 plant if Kranz leaf anatomy is observed or if two green cell types predominate the leaf structure in an unusual arrangement, as in Suaeda. However, C_4 photosynthesis can not be recognized simply by observing two green cell types in a leaf, e.g., palisade and spongy mesophyll, rather a cellular biochemical specialization must be present. To verify the cellular specialization of a C_4 plant several experimental approaches are available such as organelle ultrastructure,[1,7] differential cell or organelle fluorescence upon illumination or the addition of a fluorescently-labeled antibody against a specific cell-type protein,[12,13] differential protein[14,15] or enzyme composition[1] of the cell types, or the demonstration of a unique activity in intact or isolated cell types.[1]

Figure 2 illustrates a striking degree of organelle ultrastructural differentiation in a C_4 plant when comparing mesophyll cells with bundle sheath cells. Analysis of ultrastructure has proved useful because many C_4 plants have chloroplasts with distinctly different lamellar stacking (Figure 2) plus other structural features.[8,9,16] However, an ultrastructural difference is not a necessary component of the C_4 photosynthetic system, as for example with bermuda grass where mesophyll and bundle sheath chloroplasts both contain large grana stacks.[16] However, the striking difference in the ultrastructure of the mesophyll

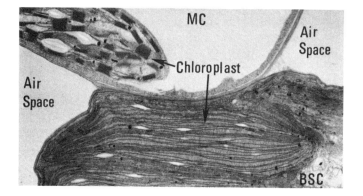

Figure 2. Electron micrograph of a Digitaria sanguinalis leaf bundle sheath cell (BSC) chloroplast and a portion of a mesophyll cell (MC) chloroplast.

and bundle sheath cell chloroplasts of crab grass, Digitaria
sanguinalis, was a most intriguing discovery. One can
imagine our excitement a dozen years ago upon realizing that
such a striking difference in chloroplast ultrastructure
existed in crab grass.[16] Inasmuch as the two cell types
could be easily isolated and separated,[17,18] an unusual
opportunity was presented to study these unique cell types.

A most informative investigation with crab grass
mesophyll and bundle sheath cells has been the comparison
of various enzyme activities in extracts from the isolated
cell types. However, before discussing the detailed biochemical differences between C_4 leaf cell types, we will ad-
dress some more fundamental questions concerning the genetic
differences between mesophyll and bundle sheath cells. As
we will illustrate, there are clear differences in the distribution of enzyme activities between mesophyll and bundle
sheath cells. One could question if these differences could
have resulted from some artifact of preparation, such as a
selective denaturation or inhibition of an enzyme activity
in one cell type and not the other. However, studies with
antibodies[13] and gel electrophoretic analysis of isolated
cell-type proteins[14 15] have completely supported the differential distribution of enzyme activities.

In the following sections we will consider the protein
and enzyme complement of C_4 leaf cell types to show that C_4
metabolism is not apportioned to CO_2 fixation only but also
includes much of a C_4 plant's biochemistry. In this work
we will find that a definite division of metabolism (or
labor) occurs between cell types and that a cooperation of
cellular activities results in the efficient operation of
various metabolic pathways in leaves of C_4 plants.

PROTEIN COMPLEMENT OF C_4 CELL TYPES

To more clearly establish that C_4 leaf cell types have
the same genetic complement and that the differences
observed in enzyme activities result from differential gene
expression, protein composition and synthesis in individual
cell types was studied. With the development by O'Farrell
of a two-dimensional electrophoretic system for separating
polypeptides,[19] it was possible to examine the entire protein population of a cell. This gel electrophoretic system
is based on separation of proteins by isoelectric focusing

in the first direction and then by molecular weights of the
sodium-dodecyl-sulfate-denatured polypeptides of these
proteins in the second direction. We have used the two-
dimensional gel system to study protein composition and differential gene expression in C_4 mesophyll and bundle sheath
cells or strands.[15,20,21] The gels shown in Figure 3
represent the polypeptides synthesized by mesophyll cells
and bundle sheath strands (strands are bundle sheath cells
still attached to the vascular bundle) when these intact
cells were fed ^{35}S via the plant transpiration stream.[15]
These results demonstrate that both cell types are capable
of de novo protein synthesis and that the complement of
proteins synthesized are not the same for the two cell
types. More detailed studies of the polypeptide composi-
tions were done using extracts of the isolated cell types
and staining the gels for protein using Coomassie blue.[20]

A comparative analysis of the labeled or stained peptides in many gels, as in Figure 3, shows that different as
well as similar proteins are present in the two cell types.
We analyzed the two-dimensional protein patterns of various
crab grass mesophyll cells and bundle sheath cells preparations by overlying an individual gel with a frosted acetate
sheet and outlining the spot for each peptide. Then the
acetate sheet was layered over other gels from both mesophyll and bundle sheath cells. From such analyses we
obtained a composite of mesophyll and bundle sheath polypeptides as shown in Figure 4. Several hundred polypeptides were analyzed in these studies. Of approximately
200 peptides found in the cells, only 47% were present in
both cell types, 36% were present only in bundle sheath
cells, and 17% were present only in mesophyll cells.[20] In
addition, we examined a species from the other two C_4 subgroups (see later discussion plus actual groupings and
lists of plant species in Tables 3 and 6) and also found a
similar but unique polypeptide distribution in their
mesophyll and bundle sheath cells. Some of the polypeptides
are known enzymes or structural proteins but others have not
been identified.[15 20]

The conclusions we reach are that less than 50% of the
proteins in the two green leaf cell types are identical and
more than half of their leaf proteins are found in only one
of the leaf cell types with the bundle sheath cell having
about twice as many unique proteins as the mesophyll cell.

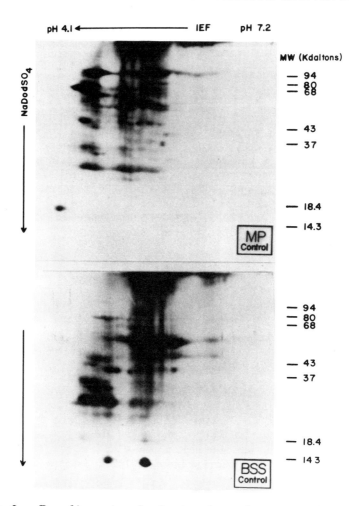

Figure 3. Two-dimensional electrophoretic separation and fluorographic detection of (^{35}S) pulse-labeled polypeptides in crab grass mesophyll cells and bundle sheath cells. Isolated cell types (mesophyll protoplasts, MP, and bundle sheath strands, BSS) from labeled crab grass seedlings were prepared for isoelectric focusing (IEF), the first dimension of separation. Samples loaded on the IEF gels were: MP control, 539 µg, 25000 cpm; BSS control, 382 µg, 26500 cpm. Second dimension gels were SDS-linear gradient (10 to 17%) polyacrylamide gels. Following completion of the second dimension, gels were stained and prepared for fluorography. NaDodSO$_4$ = sodium dodecyl sulfate.

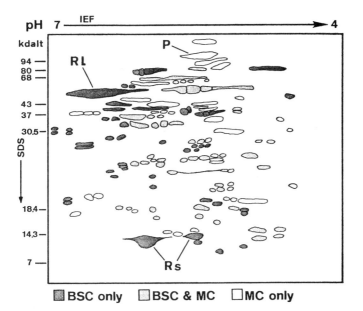

Figure 4. Composite map of bundle sheath and mesophyll proteins from crab grass. This map was made by comparing several replicate gels at different protein concentrations for both cell types. Shading indicates intercellular localization of proteins. RuBP carboxylase large subunit (R1) and small subunit (Rs), and PEP carboxylase (P) are identified.

These results are compatible with other C_4 plant work demonstrating that the DNA from mesophyll and bundle sheath chloroplasts of <u>Panicum maximum</u> are identical[22] and with the demonstration of different mRNA populations in the two cell types of <u>Zea mays</u> leaves.[23] These data all point in the direction of showing that the mesophyll and bundle sheath cell have identical DNA and that the differences observed in polypeptide composition are a result of differential expression. These conclusions are, of course, in line with the demonstrated totipotency of plant cells explanted from fully differentiated tissues. Thus the protein complement of mesophyll and bundle sheath cells demonstrates differential gene expression with the resultant cellular

specialization and cooperation known to exist in the metabolic reactions of C_4 plants.

CARBON METABOLISM

Certainly recognition of the formation of C_4-dicarboxy-
lic acids as the initial products of CO_2 fixation[3,4] ignited
the search for a new photosynthetic pathway of carbon assimilation, but also out of that pioneering work emerged the
enhanced appreciation of cellular specialization in plant
metabolism which has now been extended into many aspects
of plant biochemistry and physiology.

Table 2 summarizes data on differential enzyme activities in isolated mesophyll and bundle cells of C_4 plants.
Similar data on other C_4 plants are available[8,24,25] and
all cases show a compartmentalization of enzyme activity.
The C_4 mesophyll cell can carboxylate CO_2 via PEP carboxylase to form oxaloacetate which either is reduced with
NADPH to form malate or transaminated with alanine (in most
cases) to form aspartate. Under steady-state conditions of
photosynthesis, the carboxylation substrate for the mesophyll cell, namely PEP, is synthesized directly from pyruvate via the action of pyruvate, P_i dikinase with a net
consumption of two ATPs per PEP formed. The pyruvate supply
for a mesophyll carboxylation reaction comes from the bundle
sheath cells in a steady stream as a byproduct of the decarboxylation reaction. The C_4 acid formed in the mesophyll
cell is apparently transported to the other C_4-cell type,
generally the bundle sheath cell (see Table 1), in a diffusion controlled process. The freshly arrived C_4 acid is
decarboxylated in the bundle sheath cell yielding CO_2 and
a 3-carbon fragment. The CO_2 is then refixed via RuBP carboxylase to yield 3-PGA and all the normal C_3 cycle photosynthetic products. The chemical nature of the 3-carbon
fragment depends on which C_4 acid is decarboxylated but it
is either pyruvate or PEP (Table 3). Diffusion controlled
transport of the 3-carbon fragment from the bundle sheath
cell to mesophyll cell to regenerate PEP would complete the
cycle of C_4 photosynthesis.

Enzyme compartmentation in C_4 photosynthesis is evident
(Table 2) and from such studies along with a variety of
other work, primarily with intact leaves or isolated cell
types, the C_4 pathway of CO_2 assimilation was formulated

about a decade ago. Figure 5 is a current scheme for photosynthetic carbon metabolism in C_4 photosynthesis. In contrast with the cellular specialization in the CO_2 portion
of C_4 metabolism just discussed, the synthesis of starch
occurs in both cell types (Figure 2, Table 2). Indeed, a
variety of enzymes are found in both cell types (Table 2)
and similar data could be presented for other portions of
metabolism such as the photosynthetic photosystems or protein synthesis. We conclude that the data in Table 2 and
Figure 5 strongly support that discussed earlier showing
differential gene expression and polypeptide composition
in C_4 leaf cell types.

Today we not only recognize that cellular specialization is found in leaves of a C_4 plant such as crab grass but
we also recognize species specialization in C_4 photosynthetic CO_2 fixation. Indeed, C_4 plants can be subdivided
into three groups based on the major decarboxylase found in
their bundle sheath cell (Table 3). The decarboxylation of
organic acids in bundle sheath cells is a key feature of C_4
carbon metabolism and the decarboxylation reaction changes
the biochemistry of both mesophyll and bundle sheath cells
(Table 3, more detailed discussions of these metabolic variations are in the current literature and in reviews).[1, 24-26]

Aspects of species specialization at the decarboxylase
level in C_4 plants are incorporated into Figure 5. But an
additional cellular specialization also is found in CO_2
release via photorespiration in that the formation of phosphoglycolate occurs in the bundle sheath cells since RuBP
carboxylase/oxygenase is only in that green C_4-cell type
(Table 2). Indeed most of the enzymes of photorespiration
are concentrated in bundle sheath cells and glycolytic
enzymes are also enhanced (Table 2); but there is reasonable
activity for these metabolic processes in both cell types.
Ultrastructural studies of C_4 leaf cell types both on
organelle distribution and immunochemical localization of
enzymes, are in agreement with these enzyme activity distributions. Thus, bundle sheath cells have been shown to be
richer in mitochondria and peroxisomes and to contain all
the immunologically reactive RuBP carboxylase. Mesophyll
cells contain the immunologically reactive PEP carboxylase
and apparently non-specialized peroxisomes, which differ
from the typical specialized peroxisomes of the bundle

Table 2. Differential enzyme activities in isolated meso-
phyll and bundle sheath cells or strands of the C_4 plants:
Digitaria sanguinalis (A), Cyperus rotundus (B) and
Digitaria decumbens (C) (N.D. = not detectable).

Enzyme	Mesophyll cell			Bundle sheath cell		
	A	B	C	A	B	C
	μmoles/mg chlorophyll/hour					
Enzymes common to both mesophyll and bundle sheath cells						
Glyceraldehyde 3-P DHase (NADP)	206	570	-	284	1,300	-
Adenylate kinase	1,900	3,100	-	1,500	1,450	-
Pyrophosphatase	1,820	1,780	-	2,660	2,480	-
Gutamine synthetase	33	-	-	47	-	-
Glutamate-OAA trans- aminase	202	-	-	148	-	-
Aspartate aminotrans- ferase	-	228	-	-	240	-
Alanine aminotrans- ferase	-	108	-	-	195	-
ADPglucose trans- glucosylase	-	390	-	-	240	-
UDPglucose trans- glucosylase	-	295	-	-	300	-
Sucrose 6-P synthase	-	200	-	-	185	-
Sucrose synthase	-	320	-	-	195	-
Phosphorylase	-	6	-	-	15	-
Enzymes compartment- alized in bundle sheath cells						
Ribulose bisP carboxylase	24	5	16	450	523	560
Ribulose 5-P kinase	76	20	-	1,434	3,820	-
Ribose 5-P isomerase	48	-	-	970	-	-
NADP malic enzyme	42	8	22	845	322	1,000

Table 2. (Continued)

Enzyme	Mesophyll cell			Bundle sheath cell		
	A	B	C	A	B	C
	μmoles/mg chlorophyll/hour					
Fructose bisP aldolase	50	44	-	558	440	-
Cytochrome c oxidase	4	2	-	45	13	-
Glycolate oxidase	2	2	-	11	9	-
Hydroxypyruvate (31	-	-	169	-	-)	
reductase (18	ND	-	109	84	-)	
Enzymes compartment- alized in mesophyll cells						
PEP carboxylase	1,220	2,100	626	22	27	22
Malic DHase (NADP)	200	600	-	ND	ND	-
Pyruvate Pi dikinase	-	230	-	-	7	-

sheath cells (see A. Huang this volume for a discussion of peroxisome specialization).

Acceptance of the scheme in Figure 5 raises questions concerning the functions, or roles, or results of two inter-connecting cycles of CO_2 metabolism spatially separated into two cell types. The environmental physiology of C_4 plants is quite different from that of C_3 or CAM plants.[1,24-26] Briefly, C_4 leaf photosynthesis is near saturation at cur-rent levels of atmospheric CO_2 and C_4 photosynthesis main-tains a high rate without a detectable loss of CO_2 due to photorespiration. Apparently, cellular specialization of C_4 tissues results in the mesophyll cell effectively trap-ping all CO_2 available either from the atmosphere or from internal respiratory processes. The combination of this CO_2 trapping and the movement of C_4 organic acids to the bundle sheath cells results in release of CO_2 at the site of RuBP carboxylase in a concentrated fashion such that CO_2

Table 3. Diversity in the biochemistry of C_4 photosynthesis.

Major BSC decarboxylase (Substrate)	Energetics of decarboxylation in BSC	Major substrate moving from: MC to BSC	BSC to MC[a]	Representative plants
NADP$^+$ malic enzyme (malate)	Production of 1 NADPH/CO_2	Malate	Pyruvate	Digitaria sanguinalis, Sorghum bicolor
NAD$^+$ malic enzyme (malate)	Production of 1 NADH/CO_2	Aspartate	Alanine/pyruvate[b]	Atriplex spongiosa, Panicum bergii
PEP carboxy-kinase (oxaloacetate)	Consumption of 1 ATP/CO_2	Aspartate	PEP[b]	Panicum maximum, Sporobolus poiretii

[a] 3-PGA also moves into the mesophyll cells to support the synthesis of hexoses and starch in all three groups of plants.

[b] Nitrogen balance must be maintained, most likely via an aminotransferase-type shuttle involving alanine. PEP also may be converted to pyruvate with the formation of ATP in BSC.

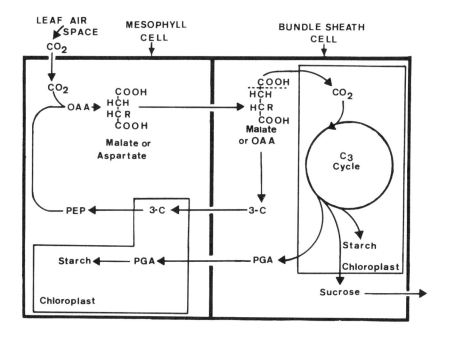

Figure 5. Current scheme for C_4 carbon metabolism showing
the cooperation between mesophyll and bundle sheath cells
in assimilating CO_2. 3-C is the 3-carbon unit remaining
after oxaloacetate or malate decarboxylation in a bundle
sheath cell.

in air is not rate limiting for C_4 photosynthesis. Hence
C_4 plants are an adaptation of photosynthesis which allows
the efficient use of current day levels of CO_2 in the
earth's atmosphere. Additional advantages accrue to the
C_4 plant as a result of its highly efficient CO_2 scavanging.
Since intracellular CO_2 concentrations can fall to quite
low levels without decreasing the fixation rate greatly,
the C_4 leaf can effectively conserve water by restricting
stomatal aperture. Thus, C_4 plants also represent an
adaptation of photosynthesis for effective utilization of
water.[24-26]

NITROGEN METABOLISM

As the cellular specialization in carbon metabolism in C_4 plants was being elucidated, we realized that some enzymes of nitrogen metabolism also exhibited cellular compartmentation.[27] Table 4 shows that the two initial enzymes in NO_3^- assimilation, namely NO_3^- reductase and NO_2^- reductase are in the C_4 mesophyll cell while enzymes for the assimilation of NH_3 are in both green cell types.[28] If we fit these enzyme activity data with the currently accepted scheme for leaf nitrogen assimilation,[29] we realize that C_4 nitrogen metabolism, like carbon metabolism, also is partially compartmentalized on a cellular basis. A scheme for C_4 nitrogen assimilation is given in Figure 6.[28] Nitrate is the major form of nitrogen available to leaves from the xylem and apparently it must pass through the bundle sheath cell on its way to the site of reduction to NH_3 in the mesophyll cell. If NH_3 or organic nitrogen is available to either mesophyll or bundle sheath cells, it can be assimilated into amino acids since both the glutamine synthetase/glutamate synthase and the glutamate dehydrogenase pathways are present in both cell types. Moreover, transaminase activities appear to be about equally distributed between the cell types. Although it has not been

Table 4.   Localization and activity of nitrogen assimilation enzymes in crab grass leaves (N.D.: not detectable).

Enzyme	Whole leaf	Mesophyll cells	Bundle sheath strands
	μmoles/mg chlorophyll/hour		
Nitrate reductase	5.1	8.3	N.D.
Nitrite reductase	25.7	39.4	N.D.
Glutamine synthetase	44.2	33.6	45.6
Glutamate synthase	10.1	6.5	13.9
Glutamate dehydrogenase			
NADH-dependent	16.7	4.1	28.3
NADPH-dependent	6.3	1.1	11.1

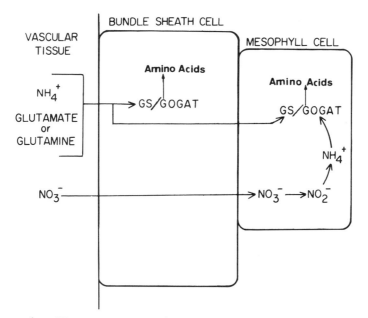

Figure 6.   Nitrogen assimilation pathways to the level of
amino acids in the two types of green cells in C_4 plants.
One pathway is for ammonium ion and organic nitrogen assimilation which occurs in both cell types.   The second pathway
is for nitrate and nitrite assimilation which occurs only
in mesophyll cells.

determined on the level of enzyme activity, we have shown
that both cell types conduct de novo protein synthesis.[15,21]

One can ask why is glutamine synthetase (GS) and glutamate synthase (GOGAT) roughly equally distributed between
the two cell types if NH_3 produced in the mesophyll cell
chloroplast is fixed into glutamate there and not allowed
to diffuse from there to the cytoplasm nor to the bundle
sheath cell.   However, NH_3 can enter the bundle sheath cell
and be fixed there via the GS/GOGAT system.   Now, we recognize that photorespiration is a bundle sheath cell process
and that GS/GOGAT are required enzymes in the reassimilation
of NH_3, which is generated in photorespiration during the
condensation of two molecules of glycine to form a serine
and CO_2.   Recent studies indicate that the reassimilation

of NH_3 takes place in the chloroplast via GS/GOGAT and the
location of these enzymes in bundle sheath cells is critical
for NH_3 reassimilation from photorespiration. But there is
a differential localization of reductive capacity, in that
both nitrate and nitrite reductase are in the mesophyll cell.
Therefore, this provides an additional electron "sink" in
the mesophyll cells so that if carbon assimilation is
limited (e.g., as in a drought) nitrogen assimilation can
pick up electrons and cycle them; thus protecting the photosystems from light inactivation. The drought sensitivity
of nitrate reductase[30] indicates that nitrogen metabolism
may not help in keeping the electron carriers turning over
under stress in the daylight. However, another form of reductive activity may be found in the recapture of photorespiratory CO_2 by the mesophyll cells, which would be the
only source of HCO_3 for PEP carboxylase when the stomata
are closed and no gases are being exchanged with the
atmosphere.

     Cellular specialization in C_4 nitrogen assimilation
also has ramifications in the intact plant. One of the most
interesting was presented by R. H. Brown who realized from
older literature data that C_4 plants, particularly grasses,
may actually utilize their available nitrogen more
efficiently than C_3 plants in producing dry matter.[31] Some
collated data of Brown's are shown in Table 5 along with
some calculations of nitrogen-use efficiency. At any
level of nitrogen fertilization the C_4 plant produces more
dry matter than the C_3 plant (Table 5). Brown theorized
that this may be due to the lower investment of nitrogen
in RuBP carboxylase by C_4 plants since they only synthesize
this enzyme in the bundle sheath cell. Indeed the RuBP
carboxylase content of C_4 leaves is less than C_3 leaves.[31]
We can conclude also from the other cellular specializations
cited in this manuscript that a variety of polypeptides are
synthesized only in one cell type including enzymes of carbon, nitrogen, and sulfur metabolism. Therefore, a division
of cellular labor occurs in the intact plant which results
in the efficient use of available nitrogen.

SULFUR METABOLISM

     Knowing that carbon and nitrogen assimilation in C_4
plants is via a cellular compartmentation of enzymes, we
extended our studies of C_4 metabolism to sulfate

Table 5. Nitrogen use efficiency, dry matter yield, and nitrogen content at various levels of nitrogen fertilization in a C_3 and a C_4 grass sod.[a]

Nitrogen applied	Dry matter (DM) yield		Nitrogen content of grass		Nitrogen use efficiency			
	metric tons/ha		% of DM		tons DM/% N		Kg forage/Kg N applied	
Kg/ha	C_3	C_4	C_3	C_4	C_3	C_4	C_3	C_4
112	3.8	8.4	2.52	2.13	1.5	3.89	33.9	74.1
224	5.8	11.4	2.77	2.26	2.09	5.04	25.9	50.9
448	7.2	16.1	3.25	2.75	2.21	5.85	16.1	35.9
896	6.9	17.5	3.5	3.0	1.97	5.83	7.7	19.5

[a]Original yield and N content data collected by R. H. Brown for fescue (C_3) and bermuda grass (C_4) from a four year field experiment on the coastal plain of Virginia.[31]

Table 6. Intercellular localization of ATP sulfurylase in the leaves of various C_4 plants. Activity of ATP sulfurylase was determined by a bioluminescence assay (N.D.: not detectable).

Organism	Whole leaf	Mesophyll protoplasts	Bundle sheath strands
$NADP^+$-malate enzyme type	μmoles/mg chlorophyll/hour		
Bothriochloa caucasica	29.3	1.6	54.3
Cymbopogon martini	22.6	2.3	88.8
Digitaria sanguinalis	42.2	6.0	91.9
Echinochloa colonum	23.6	3.8	56.0
Echniochloa crus-galli	18.4	1.4	72.6
Euchlaena mexicana	24.1	2.4	64.4
Pennisetum americanum	79.1	0.7	183.8
Sorghum bicolor	2.4	1.9	93.8
Zea mays	5.3	0.4	16.9
NAD^+malate enzyme type			
Chloris distichophylla	26.6	N.D.	36.8
Eleucine indica	16.2	0.8	21.5
Panicum bergii	24.4	0.8	39.7
Panicum miliaceum	10.9	N.D.	15.5
PEP-carboxykinase type			
Brachiaraia erucaeformis	20.0	3.2	64.3
Chloris gayana	28.7	1.1	53.2
Panicum maximum	35.5	2.5	37.8
Panicum molle	33.0	4.1	67.2
Urochloa mosambicensis	51.2	0.5	162.9

assimilation. We studied ATP sulfurylase, the initial enzyme in sulfate assimilation,[32,33] in isolated C_4 mesophyll and bundle sheath cells and sulfate assimilation in intact leaves. The cellular distribution of ATP sulfurylase for the three major subgroups of C_4 plants is given in

Table 6.[34] Surprisingly the enzyme was in the bundle sheath cell. We also found in crab grass a thiosulfonate reductase activity 2- to 3-fold higher on a protein basis in bundle sheath cells than in mesophyll cells although it was reasonably active in both cell types.[33] In addition, with intact crab grass plants fed [^{35}S]sulfate, ^{35}S-labeled proteins were found in isolated mesophyll and bundle sheath cells (Figure 3).[15]

A scheme for C_4 leaf sulfate assimilation is shown in Figure 7. As sulfate comes up through the vascular tissue it moves into the adjacent bundle sheath cell where it is activated via ATP sulfurylase. We have not assayed for the remaining enzymes in sulfur assimilation in both cell types, but we know the reductase is in both cells and the synthesis of sulfur-containing proteins occurs in both cells starting from sulfate. Therefore, the assimilation of sulfur is initiated in the bundle sheath cell of C_4 plants but

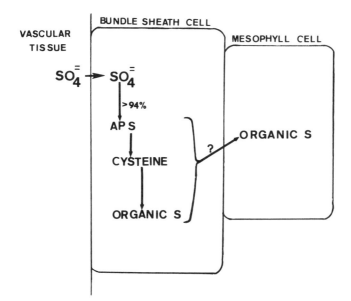

Figure 7. Sulfate assimilation pathway to the level of cysteine and organic sulfur in the two green cell types of C_4 leaves.

incorporation into other forms of organic sulfur occurs in both green cell types.

CONCLUSIONS AND PROSPECTUS

Cellular specialization and the resultant cellular co-operation is a central aspect of C_4 metabolism. A division of labor between the two green cell types found in all C_4 plants is evident in the assimilation of essential elements such as carbon, nitrogen, and sulfur. An integrative inter-pretation of current data and ideas in C_4 plants metabolism is given in Figure 8. The assimilation of carbon and nitrogen is initiated in the C_4 mesophyll cell while the assimilation of sulfur is initiated in the bundle sheath cell.

Much of the data to support this integrative scheme came from work with crab grass. No doubt this work should be extended particularly to plants with special leaf anatomy

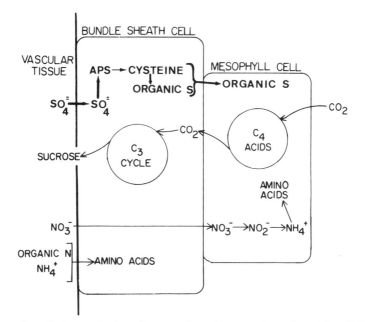

Figure 8. Integrated scheme of major routes for the inter-cellular assimilation of CO_2, nitrate, and sulfate during C_4 metabolism in crab grass leaves.

(Table 1) where other variations in C_4 metabolism likely
exist. Many other aspects of cellular differentiation and
its regulation in C_4 plants remain to be elucidated. But
clearly the compartmentation of C_4 photosynthesis into two
types of green cells shows that nature has evolved ways to
use cell types to carry out individual functions in an
efficient cooperative manner for the assimilation of
essential elements by C_4 plants.

ACKNOWLEGEMENTS

This work was supported by the National Science
Foundation Grants PCM 770-8548 (C.C.B.) and PCM-7915298
(W.H.C.).

REFERENCES

1. Ray, T. B., C. C. Black, 1979. The C_4 pathway and its
   regulation. In Photosynthesis II: Photosynthetic
   Carbon Metabolism and Related Processes (M. Gibbs,
   E. Latzko, eds.). Encyclopedia of Plant Physiology,
   New Series, Vol. 6., Spring-Verlag, Berlin.
   pp. 77-101.
2. Haberlandt, G. 1914. Physiological Plant Anatomy.
   Translation of 4th Edition. Macmillian and Co.,
   London. pp. 261-301.
3. Kortschak, H. P., C. E. Hartt, G. O. Burr. 1965.
   Carbon dioxide fixation in sugarcane leaves. Plant
   Physiol. 40: 209-213.
4. Hatch, M. D., C. R. Slack. 1966. Photosynthesis by
   sugarcane leaves. A new carboxylation reaction and
   the pathway of sugar formation. Biochem. J.
   101: 103-111.
5. Downes, R. W., J. D. Hesketh. 1968. Enhanced photosynthesis at low oxygen concentrations: differential
   response of temperate and tropical grasses. Planta
   78: 79-84.
6. Downton, W. J. S., E. B. Tregunna. 1968. Carbon
   dioxide compensation--its relation to photosynthetic
   carboxylation reactions, systematics of the graminae,
   and leaf anatomy. Can. J. Bot. 46: 207-215.
7. Johnson, H. S., M. D. Hatch. 1968. Distribution of
   the C_4-dicarboxylic acid pathway of photosynthesis
   and its occurrance in dicotyledonous plants.
   Phytochemistry 7: 375-380.

8. Laetsch, W. M. 1968. Chloroplast specialization in dicotyledons possessing the C_4-dicarboxylic acid pathway of photosynthetic CO_2 fixation. Am. J. Bot. 55: 875-883.

9. Kennedy, R. A., W. M. Laetsch. 1974. Plant species intermediate for C_3, C_4 photosynthesis. Science 184: 1097-1089.

10. Brown, R. H., W. V. Brown. 1975. Photosynthetic characteristics of Panicum milioides, a species with reduced photorespiration. Crop Sci. 15: 681-685.

11. Shomer-Ilan, A., S. Beer, Y. Waisel. 1975. Suaeda monoica, a C_4 plant without typical bundle sheaths. Plant Physiol. 56: 676-679.

12. Elkin, L., R. B. Park. 1975. Chloroplast fluoresence of C_4 plants. II. A photographic technique for obtaining relative fluorescence yields and spectra photographically. Planta 127: 187-199.

13. Hattersley, P. W., L. Watson, C. B. Osmond. 1977. In situ immunofluorescent labelling of ribulose-1,5-bisphosphate carboxylase in leaves of C_3 and C_4 plants. Aust. J. Plant Physiol. 4:523-539.

14. Huber, S. C., T. C., Hall, G. E. Edwards. 1976. Differential localization of Fraction I protein between cell types. Plant Physiol. 57: 730-733.

15. Potter, J. W., C. C. Black. 1982. Differential protein composition and gene expression in leaf mesophyll cells and bundle sheath cells of Digitaria sanguinalis (L.) Scop. Plant Physiol. In Press.

16. Black, C. C., H. H. Mollenhauer. 1971. Structure and distribution of chloroplasts and other organelles in leaves with various rates of photosynthesis. Plant Physiol. 47: 15-23.

17. Edwards, G. E., S. S. Lee, T. M. Chen, C. C. Black. 1970. Carboxylation reactions and photosynthesis of carbon compounds in isolated mesophyll and bundle sheath cells of Digitaria sanguinalis (L.) Scop. Biochem. Biophys. Res. Commun. 39: 389-395.

18. Edwards, G. E., C. C. Black. 1971. Isolation of mesophyll cells and bundle sheath cells from Digitaria sanguinalis (L.) Scop. leaves and a scanning microscopy study of the internal leaf cell morphology. Plant Physiol. 47: 149-156.

19. O'Farrell, P. H. 1971. High resolution two-dimensional electrophoresis of proteins. J. Biol. Chem. 250:4007-4021.

20. Harrison, P. A. 1980. The use of two-dimensional electrophoresis in analyzing plant proteins in cellular and subcellular systems. Ph.D. Dissertation. University of Georgia (Athens), pp. 1-154.

21. Potter, J. W. 1981. A gel electrophoretic study of protein in mesophyll cells and bundle sheath cells of a C_4-photosynthesis plant, Digitaria sanguinalis (L.) Scop. M.S. Thesis, University of Georgia (Athens), pp. 1-103.

22. Walbot, V. 1977. The dimorphic chloroplasts of the C_4 plant Panicum maximum contain identical genomes. Cell 11: 729-737.

23. Link, G., D. M. Coen, L. Bogorad. 1978. Differential expression of the gene for the large subunit of ribulose bisphosphate carboxylase in maize leaf cell types. Cell 15: 725-731.

24. Burris, R. H., C. C. Black. 1976. CO_2 Metabolism and Plant Productivity. University Park Press, Baltimore. pp. 1-431.

25. Hatch, M. D., C. B. Osmond, R. O. Slatyer. 1971. Photosynthesis and Photorespiration. Wiley-Interscience, New York. pp. 1-565.

26. Black, C. C. 1973. Photosynthetic carbon fixation in relation to net CO_2 uptake. Annu. Rev. Plant Physiol. 24: 253-286.

27. Chen, T. M., P. Dittrich, W. H. Campbell, C. C. Black. 1974. Metabolism of epidermal tissues, mesophyll cells, and bundle sheath strands resolved from mature nutsedge leaves. Arch. Biochem. Biophys. 163: 246-262.

28. Moore, R., C. C. Black. 1979. Nitrogen assimilation pathways in leaf mesophyll and bundle sheath cells of C_4 photosynthesis plants formulated from comparative studies with Digitaria sanguinalis (L.) Scop. Plant Physiol. 64: 309-313.

29. Lea, P. J., B. J. Miflin. 1974. Alternative route for nitrogen assimilation in higher plants. Nature 251: 614-616.

30. Morilla, C. A., J. S. Boyer, R. H. Hageman. 1973. Nitrate reductase activity and polyribosomal content of corn (Zea mays L.) having low leaf water potential. Plant Physiol. 51: 817-824.

31. Brown, R. H. 1978. A difference in N use efficiency in C_3 and C_4 crop plants and its implications in adaptation and evolution. Crop Sci. 18: 93-98.

32. Schwenn, J. B., A. Trebst. 1976. Photosynthetic sulfate reduction by chloroplasts. In The Intact Choloroplast (J. Barber ed.). Elsevier/North Holland Biomedical Press (The Netherlands). pp. 315-334.
33. Gerwick, B. C., C. C. Black. 1979. Sulfur assimilation in C$_4$ plants. Plant Physiol. 64: 590-593.
34. Gerwick, B. C., S. B. Ku, C. C. Black. 1980. Initiation of sulfate activation: A variation in C$_4$ photosynthesis plants. Science 209: 513-515.

Chapter Eight

THE SYNTHESIS, STORAGE AND DEGRADATION OF PLANT NATURAL
PRODUCTS: CYANOGENIC GLYCOSIDES AS AN EXAMPLE

ADRIAN J. CUTLER AND ERIC E. CONN

Department of Biochemistry and Biophysics
University of California
Davis, California 95616

INTRODUCTION

Cyanogenic glycosides are natural products found in a
wide variety of plant genera including Sorghum, Prunus,
Linum, Manihot and Passiflora.

Altogether over 2,000 species from many geographic and
climatic areas are known to be cyanogenic.[1] The structures
of some typical cyanogenic compounds are shown in Figure 1.
The aglycone moieties are derived from amino-acids, and
known cyanogenic precursors include valine, isoleucine,
leucine, tyrosine and phenylalanine.[2] When cyanogenic
tissue is crushed, free hydrogen cyanide (HCN) is usually
released. This phenomenon occurs when the tissue contains
specific β-glucosidases. The release of free HCN has led to
the suggestion that cyanogenic glycosides are defense compounds, and evidence that the presence of such products
reduces herbivore predation has been obtained by Jones.[3]

Extensive studies on the biosynthesis of cyanogenic
glycosides have been carried out both in intact plants or
plant parts and more recently in cell-free systems. This
review will summarize recent studies on the biosynthesis
of cyanogenic glycosides in cell-free systems and discuss

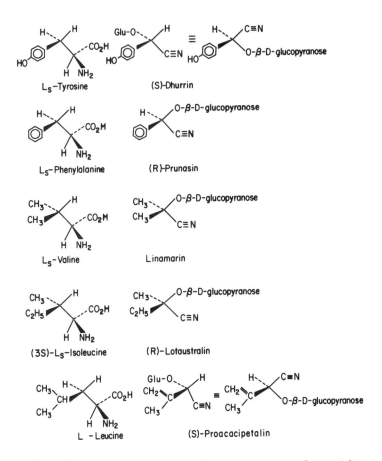

Figure 1.   Structures of typical cyanogenic glycosides and their precursor amino acids.

the way in which these compounds are stored in the leaves of <u>Sorghum</u> <u>bicolor</u>.

BIOSYNTHESIS

General Features

The biosynthesis of cyanogenic glycosides has been studied in vitro in three different systems. Most work has been performed on the grass <u>Sorghum</u> <u>bicolor</u> from which

a cell-free system was obtained which synthesized dhurrin
(Figure 1), a cyanogenic glycoside derived from tyrosine.[4,5] An analogous biosynthetic pathway has been
studied in the grass <u>Triglochin</u> <u>maritima</u> which produces
taxiphyllin, the epimer of dhurrin.[6,10] More recently the
biosynthesis of the aliphatic cyanogenic glycosides
linamarin and lotaustralin, which are derived from valine
and isoleucine respectively (Figure 1), has been studied
in <u>Linum</u> <u>usitatissimum</u> (linen flax).[7]

     All of the cell-free systems studied so far have some
features in common. For example, conversion of the amino-
acid into the corresponding cyanohydrin occurs in microsomal
particles by an oxidative reaction sequence requiring NADPH
and oxygen. Glucosylation of the cyanohydrins to form
cyanogenic glucosides is catalyzed by soluble glucosyl
transferases requiring UDPG as the glucose donor.

     Microsomal preparations have normally been obtained
from young etiolated seedlings. The tissue is homogenized
in a mortar and pestle and centrifuged at 10,000 g to remove
cellular debris. The supernatant is centrifuged at 100,000
g for 90 min to sediment the microsomal fraction. Glucosyl
transferase activity remains in the supernatant solution.
The particulate fraction is then suspended and recentrifuged
to wash away any HCN adhering to the particles (the HCN
arises from the decomposition of cyanogenic glycosides
during the initial homogenization). All operations are
carried out at pH 7.5 to 8 in Tricine or phosphate buffer
and in the presence of a reducing agent such as dithiothreitol or mercaptoethanol. The cyanohydrins formed by
the membrane-bound enzyme complexes are unstable and spon-
taneously decompose to liberate HCN and the corresponding
carbonyl compound as illustrated in Figure 2. HCN can be
detected colorimetrically[8] and a variety of assays have
been developed which use radioactive substrates to measure
the partial reactions of the different reaction sequences.[5-7]
The chemical syntheses of the radioactive intermediate for
use with the sorghum and triglochin systems were initially
developed by Møller[9] although in recent studies quicker
synthetic methods having lower yields have been employed.[6]
Although etiolated tissue is routinely used for preparing
the microsomal fractions, activity has also been obtained
from fully green seedlings of triglochin and flax. It is
worth noting that no enzymic activity is obtained in

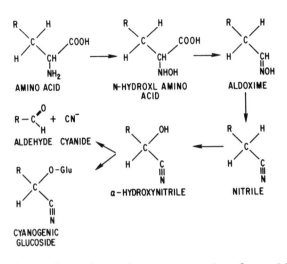

Figure 2.   General pathway for cyanogenic glycoside biosynthesis. Reproduced from Ref. 2 with permission.

triglochin or flax if the initial homogenization is performed in the presence of seed-coats.[6,7]

A general pathway for the production of cyanogenic
glycosides which is based on studies carried out to date is
illustrated in Figure 2. The first step is performed by a
monooxygenase utilizing oxygen and NADPH to produce an
N-hydroxyamino-acid. Inhibitor studies in sorghum have
indicated that this step does not involve cytochrome P-450.[5]
N-Hydroxylation is an unusual type of enzymic reaction and
few other examples are known. N-Hydroxyamino-acids have
been postulated as intermediates in the biosynthesis of
hadacidin[11] and glucosinolates[12] although conclusive
evidence is lacking. On the other hand, the N-hydroxylation of a variety of amine drugs has been demonstrated in
several animal species.[13] The second step in the general
sequence is a complex oxidative decarboxylation producing an
aldoxime. Little is known of this reaction, especially with
regard to any oxidative cofactors involved. Møller and Conn
proposed 3-(p-hydroxyphenyl)-2-nitrosopropionic acid as an
intermediate in this oxidative decarboxylation;[5] unfortun-

ately nitroso compounds are notoriously labile and therefore
it has been impossible to synthesize and test such compounds
as substrates in the usual way. The dehydration of the
aldoxime to the nitrile appears to be the simplest reaction
in the sequence but in both sorghum and triglochin the
reducing agent NADPH is essential.[6,14] The final step
catalyzed by the membrane-bound enzyme system is a stereo-
specific hydroxylation converting the nitrile to the
cyanohydrin.[14]

In all cases observed so far the glucosylation of the
cyanohydrin to yield the corresponding cyanogenic glucoside
is performed by a soluble transferase utilizing UDPG. These
enzymes have been partially purified from sorghum,[15] where
glucose is transferred to p-hydroxy-(S)-mandelonitrile to
form dhurrin, and from flax,[16] where glucose is transferred
to acetone cyanohydrin and 2-butanone cyanohydrin to form
linamarin and lotaustralin respectively. Both enzymes
catalyze the formation of β-glucosides, exhibit high
specificity for UDPG and have pH optima in the range 8 to 9.
The sorghum enzyme is active toward both p-hydroxy-(S)-
mandelonitrile and (S)-mandelonitrile, but not the (R)-
isomers, whereas the flax enzyme is not only active toward
a variety of aliphatic cyanohydrins but exhibits no stereo-
specificity toward enantiomers of its natural substrate,
2-butanone cyanohydrin.

Given the assumptions and simplifications that are inherent in the Michaelis-Menten approach to enzyme kinetics
it is not surprising when complex multienzyme systems such
as those responsible for cyanogenic glycoside biosynthesis
exhibit complex inhibitory effects. Attention has been
previously drawn to some complicated kinetic effects in triglochin.[6] In this plant, all three of the pathway intermediates exhibited pronounced substrate inhibition. In
addition, when tyrosine was also present as a substrate its
rate of utilization was markedly reduced when the intermediates were present at higher concentrations. The conversion of p-hydroxyphenylacetonitrile to p-hydroxy-(R)-
mandelonitrile was particularly vulnerable to inhibition by
N-hydroxytyrosine or p-hydroxyphenylacetaldoxime. In flax
a marked inhibition of valine metabolism occurs in the
presence of isobutyraldoxime (unpublished data). Finally,
in sorghum it was observed that p-hydroxyphenylacetonitrile
inhibited tyrosine metabolism.[17]

Other multienzyme systems commonly exhibit similar complex kinetic effects. An example is provided by kaurene synthase from the fungus Fusarium moniliforme in which the intermediate copalyl-pyrophosphate exhibited substrate inhibition when its concentration was raised above the K_m.[34] Finally, in general terms, it is becoming clear that many monooxygenase enzymes exhibit non-Michaelian kinetics[35] and therefore, that complex kinetics are very much an inherent feature of many of the complicated multienzyme systems that are increasingly being characterized.[39]

The enzymes of sorghum show a high substrate specificity for the amino-acid metabolized and no conversion of D-tyrosine, L-phenylalanine or D- and L-histidine is detected.[5] A number of intriguing questions can be asked about the flax system since two amino-acids (valine and iso-leucine) are metabolized. There seems to be a high degree of specificity in this system also since even leucine is not detectably converted to products (unpublished results) despite the fact that it is itself a precursor of cyanogenic glycosides in other cyanogenic plants.[1,2]

There are many problems still remaining to be solved regarding the biosynthesis of cyanogenic glycosides. For example, little is known of the pathway by which triglo-chinin, the ring-cleaved tyrosine derivative is formed, and the way in which cyclopentenoid compounds such as tetraphyl-lin B are produced.[1]

Metabolic Channeling

The most striking feature of all three cell-free bio-synthetic systems is that the four step conversion of the amino-acid into the cyanohydrin occurs with negligible ac-cumulation of pathway intermediates. This observation cannot be explained on the basis of the kinetic parameters alone. In all of the systems the concentrations of the intermediates are well below the concentrations that would be expected if a steady state were established between independent soluble enzymes. Also the measured V values do not consistently increase with each intermediate in the order of their utilization in the pathway (Table 1). The latter observation is paradoxical since it implies that there are rate-limiting steps in each sequence which should lead to the build-up of intermediates. For example, in sorghum

the measurable V_m for nitrile metabolism (43 nmol/h/g fsh wt,
Table 1) is considerably less than that for tyrosine (123
nmol/h/g fsh wt, Table 1) and oxime (340 nmol/h/g fsh wt,
Table 1) metabolism, but no nitrile accumulates during
enzyme reactions initiated with either tyrosine or oxime.
It follows from this that attempts to trap intermediates
derived from the amino-acid substrate by adding the
identical intermediate exogenously will be limited by the
degree to which compounds generated in situ can equilibrate
with the exogenous material. Intermediates so limited are
referred to as being channeled. To quantitate this phenomenon we divide the percentage of conversion to products of
the intermediate generated from tyrosine by the percentage
of conversion of the added intermediate. Many of our
experiments in sorghum and triglochin used a double-label
technique with ^{3}H-labelled tyrosine and ^{14}C-labelled
intermediates.[6,17] In this case the channeling values can
be calculated from the following expression:

$$\frac{\dfrac{^3\text{H-product}}{^3\text{H-product} + \,^3\text{H-intermediate}}}{\dfrac{^{14}\text{C-product}}{^{14}\text{C-product} + \,^{14}\text{C-intermediate}}}$$

A value of 1 or more for this ratio indicates that
significant channeling occurs since the endogenous material
is being more efficiently utilized despite the fact that,
at least initially, there is a large excess of exogenous
material. Channeling values for the sorghum and
triglochin intermediates are shown in Table 1. Quite
clearly in sorghum both N-hydroxytyrosine (9.5) and
p-hydroxyphenylacetonitrile (12.9) are channeled, whereas
in triglochin it is N-hydroxytyrosine (1.9) and p-hydroxyphenylacetaldoxime (1.1) that exhibit this property.

When the channeling results from sorghum were first
obtained it was suggested that two bifunctional complexes
were involved.[17] The first one would convert an amino-acid
(L-tyrosine in sorghum) into the corresponding oxime and was
common to plants containing glucosinolates. The second
complex would convert the oxime into a cyanohydrin and was
specific to cyanogenic species. This hypothesis rested on
the assumption that a relatively unchanneled intermediate

Table 1. Channeling and V_{max} values for substrates of the sorghum and triglochin microsomal systems.[a]

Plant systems	Substrates						
	Tyrosine	N-Hydroxytyrosine		p-Hydroxyphenyl acetaldoxime		p-Hydroxyphenyl acetonitrile	
	V_m[b]	V_m[b]	Channeling[c] ratio	V_m[b]	Channeling[c] ratio	V_m[b]	Channeling[c] ratio
Sorghum bicolor	123	293	9.5	340	0.8	43	12.9
Triglochin maritima	19	49	1.9	47	1.1	224	0.47

[a] Data were obtained from references 5 and 6.

[b] Expressed as nmoles HCN/h/g fresh weight.

[c] Channeling ratios calculated as described in the text and in reference 6.

passes between active sites on two separate enzyme complexes, whereas a channeled intermediate passes between active sites in the same protein. Additional support for this idea was obtained by experiments in which microsomal preparations from sorghum were prepared in the absence of mercaptoethanol and dialyzed in aerated buffer overnight.[4,5] Under these conditions the microsomal extracts rapidly converted tyrosine into p-hydroxyphenylacetaldoxime but almost no p-hydroxybenzaldehyde or p-hydroxyphenylacetonitrile accumulated. The second half of this multienzyme sequence is much more labile when a disulfide reducing agent is absent. The results from triglochin do not support this idea and in addition, there is no evidence to suggest a channeling pattern similar to that of sorghum exists in the flax enzymes (unpublished results). We have speculated elsewhere[6] that the arrangement of the pathway in triglochin may be related to the presence of a second cyanogenic glucoside, triglochinin, for which p-hydroxyphenylacetonitrile may be a direct precursor.

The relatively low channeling value for p-hydroxyphenylacetaldoxime (0.80) in sorghum does not necessarily mean that it is completely unchanneled. In assays of the sorghum enzymes using fresh microsomal preparations the oxime appears in only minute quantities[4,5,17] whereas p-hydroxyphenylacetonitrile in triglochin accumulates significantly[6] and also has a lower channeling value (0.47). The sorghum oxime seems to be an example of a partially channeled intermediate. None of the intermediates listed in Table 1 can be detected in intact sorghum seedlings after labelling studies with ^{14}C-tyrosine. Interestingly, there is also no evidence to suggest the formation of free p-hydroxymandelonitrile in vivo.[47] We interpret this observation as indicating that the cyanohydrin is channeled in the intact cell. The molecular structure responsible for such channeling is destroyed during the preparation of cell-free extracts and cross-linking studies failed to show any association between the glucosyl transferase and the microsomal particles.[17] It may be that channeling is a more widespread phenomenon in intact cells than experiments from cell-free extracts would seem to indicate.

Studies of the channeling of valine conversion to acetone cyanohydrin in flax indicate that all three inter-

mediates (N-hydroxyvaline, isobutyraldoxime and isobutyroni-
trile) are channeled (unpublished results).

An increasing number of examples of metabolic channeling in vitro appear in the literature. The well-known
and much studied fatty-acid synthetase of yeast is an
excellent example of a highly channeled enzyme complex.[19]
In this case channeling is accomplished by the covalent
binding of fatty acids of intermediate chain length to the
acyl carrier protein. Studies by Kindl have shown the
association of membrane-bound phenylalanine ammonia-lyase
and the o- and p-hydroxylases of cinnamic acid in channeled
complexes from a variety of tissues including cucumber
cotyledons,[20] potato tubers[21] and the green alga Dunaliella
marina.[22] Another good example is provided by tryptophan
synthase in Neurospora in which Matchett showed by isotope
dilution experiments that indole was highly channeled.[23]
Recently, Mally et al. presented evidence of channeling in
the pyrimidine biosynthesis of Syrian hamster cells.[24] They
studied a multifunctional enzyme responsible for the conversion of carbamyl phosphate to dihydroorotate in which
carbamyl aspartate was the channeled intermediate. The
channeling phenomenon has also been observed in the DNA
metabolism of Chinese hamster embryo fibroblast cells[25] and
in the multienzyme sequence responsible for the degradation
of cyclic AMP in beef adrenal cortex.[26] This is a far from
exhaustive list but in all cases, although a variety of
experimental approaches have been used, the preferential
metabolism of intermediates produced in situ over those
added externally is observed.

Several years ago Gaertner et al. defined a property
called catalytic facilitation with respect to the enzyme
complexes involved in the biosynthesis of aromatic amino-
acids in Neurospora.[27] This included the so-called "arom"
conjugate enzymes. According to these workers catalytic
facilitation occurred when "The rates of the over-all
sequential reactions catalyzed by the two complexes were
found to be greater than the rates of reactions initiated
later in the sequence with known intermediates.". As can
be seen from the rates shown in Table 1, the biosynthetic
pathways in sorghum and triglochin do not exhibit the phenomenon of catalytic facilitation. For example, in sorghum
the V_m for N-hydroxytyrosine (293 nmol/h/g fresh wt) is
faster than that of tyrosine (123 nmol/h/g fresh wt), and

in triglochin both of the channeled intermediates
(N-hydroxytyrosine and the oxime) are metabolized faster
than tyrosine. In more recent publications Gaertner et al.
have considered their phenomenon to be due to the coordinate
activation of the complex by the initial substrate, at least
in the case of complex I (converting 3-deoxy-D-arabino-hep-
tulosonate 7-phosphate into 3-enolpyruvylshikimate
5-phosphate), rather than due to channeling alone.[28,29,39]
Catalytic facilitation seems to be a property of this
multienzyme system which may or may not also possess the
capacity to channel intermediates.

The most likely explanation for channeling in both
soluble and membrane-bound systems is that intermediates
remain associated with an enzyme either by covalent binding
(as in fatty acid synthetase) or by being released into a
microcompartment in the protein from which, and into which,
diffusion is limited. This microcompartment also contains
the active site for the next enzyme in the pathway.

Of course, it is important to ask what the function of
channeling is and to try and understand what advantage the
cell obtains by channeling the intermediates of its metabolic pathways. One advantage that is usually cited occurs
when an intermediate is involved in more than one metabolic
pathway. In some cases channeling is demonstrated when loss
of any isozyme catalyzing the synthesis of the common intermediate leads to loss of viability. A good example of this
is provided by carbamyl phosphate in Neurospora which is
utilized in pathways leading to both arginine and uridylic
acid.[30] In this case channeling serves to maintain separate
compartments so that competing pathways are more efficiently
regulated. For cyanogenic glycosides (and other plant
natural products) this function is probably not important
because the structures of the pathway intermediates tend to
be unique in cellular metabolism.

The fact that channeled intermediates are not free to
diffuse into the surrounding medium implies that the transit
or diffusion time for transference of an intermediate to the
next enzyme in the pathway is short. Gaertner has pointed
out that this may be important in maximizing the steady
state rate, particularly in those cases when the reaction
would otherwise be diffusion limited due to the highly
viscous nature of cytoplasm.[29]

In situations where an intermediate is labile a short
transit time may also be important in preventing the accumulation of decomposition products. The N-hydroxyamino-acids
may be examples of chemically labile compounds. In
particular, N-hydroxytyrosine is known to be unstable when
the pH is slightly alkaline,[9] a condition typical of plant
cell cytoplasm. An interesting enzymatic example of this
rationale for channeling has been studied by Traut[31] in
Ehrlich ascites cells. He characterized an enzyme complex
of pyrimidine biosynthesis called complex U, responsible for
the conversion of orotate to 5'-uridylic acid, in which OMP
is the channeled intermediate. It seems that channeling
spares OMP from degradation to orotidine by a competing
nucleotidase activity. Interestingly in yeast cells, which
do not contain significant nucleotidase activity, OMP is not
channeled and orotate phosphoribosyltransferase and oroti-
dine-5'-phosphate decarboxylase are present as distinct
enzymes. With reference to the cyanogenic glycosides it
was observed several years ago that if flax seedlings were
treated in vivo with ^{14}C-valine and $\underline{D},\underline{L}$-O-methylthreonine
then ^{14}C-isobutyraldoxime-O-glucoside accumulated.[32] It
seems likely that the threonine derivative inhibited the
metabolism of isobutyraldoxime which then accumulated and
acted as a substrate for a glucosyltransferase. The experimental situation described above seems to provide another
example where channeling spares an intermediate from a
spurious reaction.

Finally, Gaertner has pointed out that channeling may
be simply a method by which the cell reduces the number of
freely-diffusing low molecular weight compounds in the cytoplasm. The presence of such compounds may tax the solvent
capacity of the cell[29] and may be important for plants such
as sorghum which synthesize large amounts of cyanogenic
glucosides in relatively short periods of time.[33]

STORAGE AND DEGRADATION

The storage and degradation of cyanogenic glycosides
has involved work on the cellular and subcellular localization of dhurrin and the enzymes involved in its metabolism
in young leaves of Sorghum bicolor. So far little work of
this type has been performed on other cyanogenic plants.

The two enzymes of Sorghum bicolor involved in the degradation of dhurrin are dhurrin β-glucosidase and hydroxynitrile lyase. Recently, two β-glucosidases catalyzing the
hydrolysis of dhurrin have been partially purified from
etiolated sorghum seedlings.[18] Both enzymes have pH optima
between 6 and 6.2 and show activity toward dhurrin and sam-
bunigrin ((S)-mandelonitrile-β-D-glucoside). Hydroxynitrile
lyase catalyzes the decomposition of p-hydroxybenzaldehyde
cyanohydrin to yield free cyanide and was partially purified
by Seely et al.[43] The enzyme is highly specific for the
(S)-isomer of p-hydroxybenzaldehyde cyanohydrin and has a
pH optimum of 5.

Free cyanide is only rapidly released from cyanogenic
glycosides in intact plants when the tissue is crushed or
otherwise disrupted. This fact implies that in the intact
tissue the glycosides and the degradative enzymes are maintained in separate compartments. The nature of the compartmentation involved was studied in Sorghum bicolor by Kojima
et al.[36] Their approach was to obtain relatively pure epidermal protoplasts, mesophyll protoplasts and bundle sheath
strands and determine the dhurrin content and enzyme
activities of each type of tissue.

To obtain mesophyll and epidermal protoplasts about 1.5
g of leaves are cut and abraded with 150-grit carborundum.
The rinsed leaves are then incubated 2 to 2.5 hours in a
buffered medium (25 mM phosphate-citrate, pH 5.5) containing
1.5% (w/v) cellulase and 0.5 M mannitol at 30°C. The leaves
are filtered through 44 μm nylon net and the released protoplasts harvested by centrifuging at 220 g for 3 min. The
pellet is resuspended and neutral red dye added to visualize
epidermal protoplasts. This compound turns bright red in
epidermal vacuoles. The protoplast suspension is then
gravity filtered through one layer of 20 μm nylon mesh which
preferentially retains the larger epidermal protoplasts.
Thus two enriched fractions are obtained, one containing
mostly mesophyll protoplasts and the other mostly epidermal
protoplasts. Cross-contamination can be further reduced by
utilizing the greater density of epidermal protoplasts.
Protoplast suspensions are allowed to stand for 15 to 30
minutes. Epidermal protoplasts sediment more rapidly and
the overlaying mesophyll enriched layer can be removed with
a Pasteur pipette. This process is repeated as necessary to
obtain pure preparations. Typically, mesophyll protoplasts

contain about 5 to 10% contamination by epidermal protoplasts and epidermal protoplasts about 10% contamination by
mesophyll protoplasts.

Bundle sheath strands are obtained by a somewhat different procedure. Leaves are cut across the veins with a
gel slicer to obtain 0.5 mm segments. The leaf segments are
then digested in a similar way to that for protoplast preparations except that a 2% solution of cellulase is used.
The incubation continues for a total of 4 hours although the
digestion medium is replaced by fresh enzymes after 2 hours.
At the end of the digestion the leaf material is allowed to
settle in a centrifuge tube and suspended cuticular fractions are removed with a Pasteur pipette. This procedure
is repeated to further purify the bundle sheath strands
which are then collected by gravity filtration through a
149-μm net.

Typical data obtained from such studies by Kojima et
al.[36] are shown in Table 2. It is immediately obvious that
dhurrin is present mostly in the epidermal protoplasts.
After correction is made for cross-contamination then 97% of
the dhurrin is in the epidermal protoplasts. On the other
hand, the data in Table 2 show clearly a large preponderance
of both dhurrin β-glucosidase and hydroxynitrile lyase in
the mesophyll protoplasts. After correction is made for
cross-contamination, only 6.5% of the glucosidase activity
and 5% of the hydroxynitrile lyase activity is found in the
epidermal protoplasts. Finally, the bundle sheath strands
contain only traces of either dhurrin or the degradative
enzymes.

In the case of sorghum, observations of thin leaf
sections showed an 8:1 preponderance of mesophyll over epidermal cells. When the results in Table 2 are corrected
for this in order to express them on a whole leaf basis
dhurrin is still present mainly in the epidermal cells (at
least 78%) but less than 1% of the β-glucosidase and
hydroxynitrile lyase activities are present in epidermal
tissue.

Kojima et al.[36] also determined the activity of
UDPG:p-hydroxybenzaldehyde cyanohydrin β-glucosyltransferase
activity (the last step of the biosynthetic pathway). They
found that approximately 70% of the total glucosyl

Table 2.   The distribution of dhurrin and enzymes involved in its metabolism in different fractions from 6-day old sorghum leaf blades.[a]

	Dhurrin (μmol)	Dhurrin β-glucosidase (μmol/hr)	Hydroxynitrile lyase (μmol/hr)
Epidermal protoplasts			
per 10⁶ protoplasts	5.3	0.55	0.99
per mg protein	26.5	2.75	4.95
Mesophyll protoplasts			
per 10⁶ protoplasts	0.47	8.5	19.7
per mg protein	1.27	23.0	53.0
per mg chlorophyll	10.4	191.0	441.0
Bundle sheath strands			
per mg protein	0.006	1.65	2.23
per mg chlorophyll	0.054	11.6	13.4

[a]Data were obtained from reference 36.

transferase activity of the leaf blade was localized in
the epidermal layer (after correction for the predominance
of mesophyll over epidermal cells). More recent experiments in this laboratory have suggested that at least 80%
of the glucosyl transferase activity is present in epidermal
cells (E. Wurtele, unpublished results). Little is known
of the localization of the remaining enzymes involved in
biosynthesis except that they are closely associated with
the endoplasmic reticulum.[42]

These results showed how dhurrin and the enzymes that
catalyze its degradation are maintained in separate compartments of the leaf. The rapid release of free cyanide which
occurs when sorghum leaves are crushed follows the rupturing
of different cell types allowing their contents to mix.
There have been few comprehensive studies of the cellular
localization of other plant secondary metabolites. There
are a few examples of natural products being sequestered in
epidermal tissue[44,45] but little is known of the relationship
of natural products and the enzymes of their metabolism.
In a recent study Oba et al.[46] found the glucosides of
2-hydroxycinnamic acids to be stored in mesophyll vacuoles
while o-coumaric acid:UDPG glucosyl transferase was present
in mesophyll cells outside the vacuole. This study also
found coumarin β-glucosidase to be extracellular. This
system provides an interesting contrast at several points
to that for dhurrin in sorghum.

Recently, we have studied the localization of dhurrin
and the enzymes involved in its metabolism at the subcellular level. Several years ago Saunders and Conn isolated
vacuoles by gentle lysis of protoplasts and purified them
by centrifugation on a discontinuous Ficoll gradient.[37] All
of the dhurrin present in protoplasts was accounted for by
that present in the vacuoles. In view of the subsequent
study by Kojima et al.[36] it seems likely that dhurrin is
located in epidermal vacuoles.

More recently Thayer and Conn considered the question
of the subcellular localization of the degradative enzymes
dhurrin β-glucosidase and hydroxynitrile lyase in sorghum
mesophyll cells.[38] Mesophyll protoplasts were prepared,
gently lysed and the cellular components fractionated by
centrifugation on a linear 30 to 55% (w/w) sucrose density
gradient. The results are shown in Figure 3. The

hydroxynitrile lyase was found in the supernatant fractions
(Figure 3C) suggesting that it is a soluble enzyme of either
the vacuole or the cytoplasm. On the other hand, dhurrin
β-glucosidase activity (dhurrinase) was associated with the
intact chloroplasts when protoplasts were prepared either
from the tip of the first leaf blade (Figure 3A) or from the
base of the second leaf blade (Figure 3B). In these experiments it was observed that the percentage recovery of triose
phosphate dehydrogenase, dhurrinase and chlorophyll put on
the gradient and collected in the 1.199 to 1.21 density
range was essentially the same. For example, 69.2% of the
dhurrinase and 64.7% of the chlorophyll in Figure 3A was
collected between the density values cited above. Thayer
and Conn also showed that the chloroplasts of mesophyll
protoplasts were stained purple by the glucosidase substrate
6-bromo-2-naphthyl-β-D̲-glucoside, a known substrate for
dhurrinase, thus confirming the chloroplast localization
of this enzyme.

Examination of Figures 3A and 3B shows a certain asym-
metry in the dhurrinase peak. In particular, a shoulder at
a density of about 1.21 is apparent, which typically represented 20 to 40% of the total chloroplast-associated dhurrinase activity. This phenomenon also appeared when protoplasts were prepared from etiolated seedlings (Figure 3E).
Examination of the dhurrinase shoulder region by light and
electron microscopy revealed a few chloroplast aggregates or
"clumps" as well as individual chloroplasts. No organelles
other than mesophyll plastids were found in this region and
Thayer and Conn could offer no simple explanation for the
dhurrinase shoulder.[36]

It is not clear from these experiments whether dhurrinase is associated with the external surface of the
chloroplast or is present internally. The absence of dhurrinase activity from broken chloroplasts suggest that either
it is a soluble protein inside the intact organelle or that
it is a soluble protein inside the intact organelle or that
it is very loosely associated by ionic interactions with a
membranous fraction.

TURNOVER

It is becoming increasingly clear that plant natural
products are often actively metabolized.[40] However, after

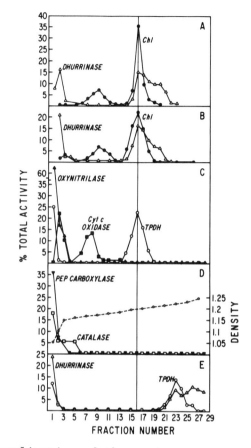

Figure 3.   Localization of the various enzyme activities and
markers in 1.35 ml fractions of 40 ml linear 30 to 55% (w/w)
sucrose density gradients of ruptured mesophyll protoplasts.
(Reproduced from Ref. 38 with permission.)   The percentage
of the total activity of each (given below the parentheses)
put on the gradient is shown.   One unit is equal to 1 μmole
product/min.   A, Chl (1.98 mg) and dhurrinase (11.2 units)
from protoplasts prepared from the tip half of the older
leaf blade; B, Chl (0.81 mg) and dhurrinase (2.73 units)
from protoplasts prepared from the basal half of the younger
leaf blade; C, TPDH (5.75 units), Cyt c oxidase (0.329
units), and hydroxynitrile lyase (2.43 units) from protoplasts prepared from etiolated leaf blades; D, phosphoenolpyruvate carboxylase (2.42 units), catalase (9.63 units),
and density of sucrose from protoplasts prepared from entire

the initial period of rapid dhurrin synthesis in green sorghum seedlings the amount of cyanide per shoot appears to
decrease only slightly for the next couple of weeks[33] (S.
Adewusi, unpublished results). A slow rate of net degradation does not necessarily mean that turnover is absent, and
indeed Bough and Gander obtained evidence through pulse
labeling experiments which indicated a turnover of about
0.05 μm of dhurrin per hour in etiolated seedlings containing 1 μmole of dhurrin per shoot.[41] In view of what we now
know of the cellular localization of the degradative enzymes
it will be interesting to know more of the importance and
mechanism of turnover in green seedlings.

SUMMARY

    These studies on the metabolism of cyanogenic glyco-
sides have revealed compartmentation at three levels. At
the tissue level in <u>Sorghum</u> <u>bicolor</u>, dhurrin is present in
epidermal cells and enzymes responsible for its degradation
are associated with mesophyll cells. At the subcellular
level, dhurrin is present in epidermal vacuoles. The
enzyme dhurrin β-glucosidase is associated with the mesophyll chloroplasts and p-hydroxymandelonitrile hydroxynitrile lyase is probably a soluble enzyme in mesophyll
vacuoles or cytoplasm. At the molecular level, the large
amounts of intermediates necessarily produced during dhurrin
accumulation in germinating seedlings remain closely
associated with the enzymes themselves in all three biosynthetic systems examined hitherto in vitro.

    These studies on the metabolism of cyanogenic glyco-
sides have already revealed many unexpected and interesting
features. It is to be hoped that this investigation with
sorghum will profice both a useful model and a basis for
comparison with the metabolism of other higher plant
natural products.

Figure 3 (legend, continued)

leaf blades; E, dhurrinase (4.24 units) and TPDH (2.63
units) from protoplasts prepared from etiolated leaf blades.
(●──●), Chl; (Δ - Δ), dhurrinase; (o──o), TPDH; (■ - ■),
Cyt c oxidase; (▲ - ▲), hydroxynitrile lyase; (▼ - ▼), PEP
carboxylase; (□ - □), catalase, and (o---o), sucrose density.

REFERENCES

1. Seigler, D. S. 1981. Recent developments in the
    chemistry and biology of cyanogenic glycosides and
    lipids. Rev. Latinoamer. Quim. 12: 39-48.
2. Conn, E. E. 1979. Biosynthesis of cyanogenic glyco-
    sides. Naturwissenschaften 66: 28-34.
3. Jones, D. A. 1979. Chemical defense: primary or
    secondary function? Am. Nat. 113: 445-451.
4. McFarlane, I. J., E. M. Lees, E. E. Conn. 1975. The
    in vitro biosynthesis of dhurrin, the cyanogenic
    glycoside of Sorghum bicolor. J. Biol. Chem.
    250: 4708-4713.
5. Møller, B. L., E. E. Conn. 1979. The biosynthesis of
    cyanogenic glucosides in higher plants. N-hydroxy-
    tyrosine as an intermediate in the biosynthesis of
    dhurrin by Sorghum bicolor (Linn) Moench. J. Biol.
    Chem. 254: 8575-8583.
6. Cutler, A. J., W. Hösel, M. Sternberg, E. E. Conn.
    1981. The in vitro biosynthesis of taxiphyllin
    and the channeling of intermediates in Triglochin
    maritima. J. Biol. Chem. 256: 4253-4258.
7. Cutler, A. J., E. E. Conn. 1981. The biosynthesis of
    cyanogenic glucosides in Linum usitatissimum (Linen
    flax) in vitro. Arch. Biochem. Biophys. 212: 468-474.
8. Lambert, J. L., J. Ramasamy, J. V. Paukstelis. 1975.
    Stable reagents for the colorimetric determination
    of cyanide by modified König reactions. Anal. Chem.
    47: 916-918.
9. Møller, B. L. 1978. Chemical synthesis of labelled
    intermediates in cyanogenic glucoside biosynthesis.
    J. Labeled Compounds Radiopharmaceuticals 14: 663-671.
10. Hösel, W., A. Nahrstedt. 1980. In vitro biosynthesis
    of the cyanogenic glucoside taxiphyllin in Triglochin
    maritima. Arch. Biochem. Biophys. 203: 753-757.
11. Stevens, R. L., T. F. Emery. 1966. The biosynthesis
    of hadacidin. Biochemistry 5: 74-81.
12. Kindl, H., E.W. Underhill. 1968. Biosynthesis of
    mustard oil glucosides: N-hydroxyphenylalanine, a
    precursor of glucotropaeolin and a substrate for the
    enzymatic and nonenzymatic formation of phenylacetal-
    dehyde oxime. Phytochemistry 7: 745-756.

13. Weisburger, J.H., E.K. Weisburger. 1973. Biochemical formation and pharmacological, toxicological, and pathological properties of hydroxylamines and hydroxamic acids. Pharmacol. Rev. 25: 1-66.

14. Shimada, M., E.E. Conn. 1977. The enzymatic conversion of p-hydroxyphenylacetaldoxime to p-hydroxymandelonitrile. Arch. Biochem. Biophys. 180: 199-207.

15. Reay, P., E.E. Conn. 1974. Purification and properties of a uridine diphosphate glucose: aldehyde cyanohydrin β-glucosyltransferase from sorghum seedlings. J. Biol. Chem. 249: 5826-5830.

16. Hahlbrock, K., E.E. Conn. 1970. The biosynthesis of cyanogenic glycosides in higher plants: Purification and properties of a uridine diphosphate glucose: ketone cyanohydrin β-glucosyltransferase from Linum usitatissimum L. J. Biol. Chem. 245: 917-922.

17. Møller, B. L., E. E. Conn. 1980. The biosynthesis of cyanogenic glucosides in higher plants. Channeling of intermediates in dhurrin biosynthesis by a microsomal system from Sorghum bicolor (Linn) Moench. J. Biol. Chem. 255: 3049-3056.

18. Eklund, S. 1981. M.S. Thesis, University of California, Davis.

19. Stoops, J. K., S. J. Wakil. 1981. Animal fatty acid synthetase: A novel arrangement of the β-ketoacyl synthetase sites comprising domains of the two subunits. J. Biol. Chem. 256: 5128-5133.

20. Czichi, U., H. Kindl. 1971. Phenylalanine ammonia lyase and cinnamic acid hydroxylases as assembled consecutive enzymes on microsomal membranes of cucumber cotyledons: Cooperation and subcellular distribution. Planta 134: 133-143.

21. Czichi, U., H. Kindl. 1975. Formation of p-coumaric acid and o-coumaric acid from L-phenylalanine by microsomal membrane fractions from potato: Evidence of membrane-bound enzyme complexes. Planta 125: 115-125.

22. Czichi, U., H. Kindl. 1975. A model of closely assembled consecutive enzymes on membranes: Formation of hydroxycinnamic acids from L-phenylalanine on thylakoids of Dunaliella marina. Hoppe-Seyler's Z. Physiol. Chem. 356: 475-485.

23. Matchett, W. H. 1974. Indole channeling by tryptophan synthase of Neurospora. J. Biol. Chem. 249: 4041-4049.

24. Mally, M. I., D. R. Grayson, R. E. Evans. 1980. Cata-
    lytic synergy in the multifunctional protein that
    initiates pyrimidine biosynthesis in Syrian hamster
    cells. J. Biol. Chem. 255: 11372-11380.
25. Reddy, G. P. V., A. B. Pardee. 1980. Multienzyme
    complex for metabolic channeling in mammalian DNA
    replication. Proc. Nat. Acad. Sci. U.S.A.
    77: 3312-3316.
26. Wombacher, H. 1980. Evidence for a membrane-bound
    multienzyme sequence degrading cyclic adenosine
    3':5'-monophosphate. Arch. Biochem. Biophys.
    201: 8-19.
27. Gaertner, F. H., M. C. Ericson, J. A. DeMoss. 1970.
    Catalytic facilitation in vitro by two multienzyme
    complexes from Neurospora crassa. J. Biol. Chem.
    245: 595-600.
28. Welch, G. R., F. H. Gaertner. 1976. Coordinate acti-
    vation of a multienzyme complex by the first sub-
    strate. Evidence for a novel regulatory mechanism
    in the polyaromatic pathway of Neurospora crassa.
    Arch. Biochem. Biophys. 172: 476-489.
29. Gaertner, F. H. 1978. Unique catalytic properties of
    enzyme clusters. Trends Biochem. Sci. 3: 63-65.
30. Davis, R. H. 1972. Metabolite distribution in cells.
    Two carbamyl phosphate gradients and their sources
    can be discerned in Neurospora. Science 178: 835-840.
31. Traut, T. W. 1980. Significance of the enzyme complex
    that synthesizes UMP in Ehrlich ascites cell. Arch.
    Biochem. Biophys. 200: 590-594.
32. Tapper, B. A., G. W. Butler. 1972. Intermediates in
    the biosynthesis of linamarin. Phytochemistry
    11: 1041-1046.
33. Stafford, H. A. 1969. Changes in phenolic compounds
    and related enzymes in young plants of Sorghum.
    Phytochemistry 8: 743-752.
34. Fall, R. R., C. A. West. 1971. Purification and
    properties of kaurene synthetase from Fusarium
    moniliforme. J. Biol. Chem. 246: 6913-6928.
35. Reilly, P. E. B., J. O'Shannessy, R. G. Duggleby.
    1980. Non-Michaelian monooxygenase kinetics:
    Studies using competitive inhibitors. FEBS Lett.
    119: 63-67.

36. Kojima, M., J. E. Poulton, S. S. Thayer, E. E. Conn.
    1979. Tissue distributions of dhurrin and of
    enzymes involved in its metabolism in leaves of
    Sorghum bicolor. Plant Physiol. 63: 1022-1028.
37. Saunders, J. A., E. E. Conn. 1978. Presence of the
    cyanogenic glucoside dhurrin in isolated vacuoles
    from Sorghum. Plant Physiol. 61: 154-157.
38. Thayer, S. S., E. E. Conn. 1981. Subcellular locali-
    zation of dhurrin β-glycosidase and hydroxynitrile
    lyase in the mesophyll cells of Sorghum leaf blades.
    Plant Physiol. 67: 617-622.
39. Welch, G. R. 1977. On the role of organized multi-
    enzyme systems in cellular metabolism: A general
    synthesis. Prog. Biophys. Molec. Biol. 32: 103-191.
40. Barz W., J. Köster. 1981. The metabolic turnover of
    secondary products. In The Biochemistry of Plants:
    Secondary Plant Products (P. K. Stumpf, E. E. Conn,
    eds.). Edit. 1, Vol. 7, Chap. 3. Academic Press,
    New York, pp. 35-84.
41. Bough, W. A., J. E. Gander. 1971. Exogenous
    L-tyrosine metabolism and dhurrin turnover in
    Sorghum seedlings. Phytochemistry 10: 67-77.
42. Saunders, J. A., E. E. Conn, C. H. Lin, M. Shimada.
    1977. Localization of cinnamic acid 4-monooxygenase
    and the membrane-bound enzyme system for dhurrin bio-
    synthesis in Sorghum seedlings. Plant Physiol.
    60: 629-634.
43. Seely, M. K., R. S. Criddle, E. E. Conn. 1966. The
    metabolism of aromatic compounds in higher plants:
    On the requirement of hydroxynitrile lyase for
    flavin. J. Biol. Chem. 241: 4457-4462.
44. Steinitz, B., H. Bergfield. 1977. Pattern formation
    underlying phytochrome-mediated anthocyanin synthesis
    in the cotyledons of Sinapis alba L. Planta
    133: 229-235.
45. Tissut, M. 1974. Etude de la localisation et dosage
    in vivo des flavonols de l'oignon. CR Acad. Sci.
    Paris Ser D 279: 659-662.
46. Oba, K., E. E. Conn, H. Canut, A. M. Boudet. 1981.
    Subcellular localization of 2-(β-D-glycosyloxy) cin-
    namic acids and the related β-glucosidases in the
    leaves of Melilotus alba DESR. Plant Physiol.
    In press.
47. Conn, E. E. 1973. Biosynthesis of cyanogenic glyco-
    sides. Biochem. Soc. Symp. 38: 277-302.